2/00

 St. Louis Community College

Forest Park
Florissant Valley
Meramec

Instructional Resources
St. Louis, Missouri

Genetic Engineering

GENETIC ENGINEERING

A Documentary History

Edited by THOMAS A. SHANNON

Primary Documents in American History and Contemporary Issues

GREENWOOD PRESS
Westport, Connecticut • London

Library of Congress Cataloging-in-Publication Data

Shannon, Thomas A. (Thomas Anthony), 1940–
    Genetic engineering : a documentary history / edited by Thomas A.
Shannon.
        p.   cm.—(Primary documents in American history and
contemporary issues series, ISSN 1069–5605)
    Includes bibliographical references (p.   ) and index.
    ISBN 0–313–30457–2 (alk. paper)
    1. Genetic engineering—History—Sources.   I. Title.   II. Series.
QH442.S476   1999
660.6'5'09—dc21   98–46808

British Library Cataloguing in Publication Data is available.

Library of Congress Catalog Card Number: 98–46808
ISBN: 0–313–30457–2
ISSN: 1069–5605

First published in 1999

Greenwood Press, 88 Post Road West, Westport, CT 06881
An imprint of Greenwood Publishing Group, Inc.
www.greenwood.com

Printed in the United States of America

The paper used in this book complies with the
Permanent Paper Standard issued by the National
Information Standards Organization (Z39.48–1984).

10 9 8 7 6 5 4 3 2 1

Copyright Acknowledgments

Contents

Series Foreword   xix

Significant Dates in the History of Genetics   xxi

Introduction   xxiii

PART I: The Debates Surrounding Genetic Engineering   1

Document   1: James Watson, *The Double Helix* (1968)   3

Document   2: J. D. Watson and F. H. Crick, "Molecular
               Structure of Nucleic Acids" (1953)   3

Document   3: Robert Sinsheimer, "Troubled Dawn for
               Genetic Engineering" (1975)   4

Document   4: Paul Berg, D. Baltimore, S. Brenner, R. Roblin,
               and M. Singer, "Asilomar Conference on
               Recombinant DNA Molecules (1975)   8

Document   5: Daniel Callahan, "Recombinant DNA: Science
               and the Public" (1977)   10

Document   6: National Institutes of Health Guidelines for
               rDNA Research (1976)   16

Document   7: Nicholas Wade, "Genetics: Conference Sets
               Strict Controls to Replace Moratorium" (1975)   18

Document   8: Allen Silverstone, Testimony to Cambridge,
               Mass., City Council with Respect to the

                    Establishment of Guidelines for rDNA
                    Research in Cambridge, Mass. (1976)   19

Document   9:   Stanley N. Cohen, "Recombinant DNA: Fact
                    and Fiction" (1977)   20

Document   10:   Jeremy Rifkin, "DNA: Have the Corporations
                    Already Grabbed Control of New Life
                    Forms?" (1977)   26

Document   11:   James S. Hall, "James Watson and the Search
                    for Biology's 'Holy Grail' " (1990)   27

PART II: Animal Applications   37

Document   12:   Rudolf Jaenisch, "Transgenic Animals" (1988)   39

Document   13:   John Travis, "Scoring a Technical Knockout in
                    Mice" (1992)   41

Document   14:   Novo Nordisk Research Corporation,
                    Statement on the Use of Transgenic Animals
                    in Research (1994)   43

Document   15:   Peter R. Wills, "Transgenics a Threat to
                    Nature" (1995)   44

Document   16:   Gary L. Francione, "Animal Rights and
                    Animal Welfare: Five Frequently Asked
                    Questions" (1996)   45

Document   17:   Philip Leder and Timothy A. Stewart,
                    Transgenic Non-Human Mammals (1988)   46

Document   18:   Robert Wachbroit, "Eight Worries about
                    Patenting Animals" (1988)   47

Document   19:   Mark Sagoff, "Animals as Inventions:
                    Biotechnology and Intellectual Property
                    Rights" (1996)   51

Document   20:   Research and Production Issues: National
                    Pork Producers Council Position Statement
                    (1997)   52

Document   21:   Alan R. Shuldiner, M.D., "Transgenic
                    Animals" (1996)   53

Document   22:   Baruch Brody, "On Patenting Transgenic
                    Animals" (1995)   56

PART III: Agriculture   61

Document 23:   Keith Schneider, "Wisconsin Temporarily
                Banning Gene-Engineered Drug for Cows"
                (1990)   63

Document 24:   Peter Wills, "Biologist Rejects Transgenic
                Foods" (1995)   63

Document 25:   Molly O'Neill, "Geneticists' Latest Discovery:
                Public Fear of 'Frankenfood' " (1992)   64

Document 26:   David Kessler, Michael R. Taylor, James H.
                Maryanski, Eric L. Flamm, and Linda S. Kahl,
                "The Safety of Food Developed by
                Biotechnology" (1992)   65

Document 27:   U.S. Food and Drug Administration,
                "Biotechnology of Food" (1994)   71

Document 28:   U.S. Food and Drug Administration Office of
                Premarket Approval, "Guidance on
                Consultation Procedures: Foods Derived from
                New Plant Varieties" (1997)   72

Document 29:   Henry L. Miller, "A Need to Reinvent
                Biotechnology Regulation at the EPA" (1994)   73

Document 30:   Anne Simon Moffat, "High-Tech Plants Promise
                a Bumper Crop of New Products" (1992)   77

Document 31:   David L. Wheeler, "The Search for More
                Productive Rice" (1995)   78

Document 32:   Barnaby J. Feder, "Out of the Lab, A
                Revolution on the Farm" (1996)   82

Document 33:   Convention on Biodiversity (1992)   83

Document 34:   Youssef M. Ibrahim, "Genetic Soybeans Alarm
                Europeans" (1996)   84

Document 35:   Jane Rissler, *Perils amidst the Promise: Ecological
                Risks of Transgenic Crops in a Global Market* (1993)   87

Document 36:   *Asgrow Seed Co. v. Winterboer* (1995)   91

PART IV: The Human Genome Project   97

Document 37:   Helen Donis-Keller and others, "A Genetic
               Linkage Map of the Human Genome" (1987)   99

Document 38:   Renato Dulbecco, "A Turning Point in Cancer
               Research: Sequencing the Human Genome"
               (1986)   100

Document 39:   Report of the National Research Council
               Committee on Mapping and Sequencing the
               Human Genome (1988)   100

Document 40:   U.S. Department of Energy, "Understanding
               Our Genetic Inheritance" (1996)   102

Document 41:   Carol A. Tauer, "The Human Significance of
               the Genome Project" (1992)   103

Document 42:   Daniel E. Koshland, Jr., "Sequences and
               Consequences of the Human Genome" (1989)   110

Document 43:   *Tarasoff v. Regents of the University of California*
               (1976)   111

Document 44:   Mark A. Rothstein, "Genetic Screening in
               Employment: Some Legal, Ethical, and
               Societal Issues" (1990)   112

Document 45:   National Institutes of Health Workshop
               Statement: "Reproductive Genetic Testing:
               Impact on Women" (1991)   112

Document 46:   National Institutes of Health–Department of
               Energy Ethics, Law, and Social Issues
               Working Group and National Action Plan
               (1995)   113

Document 47:   The Genetic Information Nondiscrimination in
               Health Insurance Act of 1997 (Proposed)   114

Document 48:   The Genetic Confidentiality and
               Nondiscrimination Act of 1997 (Proposed)   115

Document 49:   The Genetic Justice Act (Proposed) (1997)   115

Document 50:   Nachama L. Wilker and others, "DNA Data
               Banking and the Public Interest" (1992)   116

Document 51:   Ted F. Peters and Robert J. Russell, "The Human
               Genome Project: What Questions Does It Raise
               for Theology and Ethics?" (1992)   117

Contents xiii

Document 52: Philip Hefner, "The Evolution of the Created Co-Creator" (1989)   121

Document 53: Jeremy Rifkin, *Algeny: A New Word—A New World* (1984)   122

Document 54: Ron Cole-Turner, *The New Genesis: Theology and the Genetic Revolution* (1993)   123

Document 55: Statement from the National Council of Churches of Christ (1980)   123

Document 56: United Methodist Church Genetic Science Task Force, Draft Report to Annual and Central Conferences (1990)   124

Document 57: United Church of Christ Statement to the Seventeenth General Synod (1989)   125

Document 58: Episcopal Church Statement to the Seventieth General Convention (1991)   125

Document 59: United Methodist Church Science Task Force, Report to the 1992 General Conference   126

Document 60: World Council of Churches, *Biotechnology* (1989)   126

Document 61: *Donum Vitae* (1987)   127

Document 62: National Human Genome Research Institute, Policy on Availability and Patenting of Human Genomic DNA Sequence (1996)   128

Document 63: Ted F. Peters, "Genome Project Forces New Look at Ethics, Law" (1993)   130

Document 64: W. French Anderson, "Genetics and Human Malleability" (1990)   136

Document 65: World Council of Churches, *Manipulating Life: Ethical Issues in Genetic Engineering* (1982)   137

Document 66: Catholic Health Association, *Human Genetics* (1990)   137

Document 67: James D. Watson and Robert M. Cook-Deegan, "Origins of the Human Genome Project" (1991)   138

Document 68: James D. Watson and Robert M. Cook-Deegan, "The Human Genome Project and International Health" (1990)   140

PART V: Issues in Research   147

Document 69: Thomas H. Murray, "Ethical Issues in Human
              Genome Research" (1991)   149

Document 70: American Society of Human Genetics,
              "Statement on Cystic Fibrosis Screening" (1989)   150

Document 71: American Society of Human Genetics,
              "Statement on Genetic Testing for Breast and
              Ovarian Cancer Predisposition" (1994)   151

Document 72: J.B.S. Haldane, *Heredity and Politics* (1938)   151

Document 73: U.S. Congress, Office of Technology
              Assessment, *The Role of Genetic Testing in the
              Prevention of Occupational Disease* (1983)   152

Document 74: The Office of Technology Assessment's Report
              on Genetic Monitoring and Screening in the
              Workplace (1991)   154

Document 75: Council for Responsible Genetics, Position
              Paper on Genetic Discrimination (1997)   160

Document 76: Nelson A. Wivel and LeRoy Walters, "Germ-
              Line Gene Modification and Disease
              Prevention: Some Medical and Ethical
              Perspectives" (1993)   161

Document 77: American Society of Human Genetics, Ad Hoc
              Statement on Insurance Issues in Genetic
              Testing (1995)   166

Document 78: President's Commission for the Study of
              Ethical Problems in Medicine and Biomedical
              and Behavioral Research, *Splicing Life: The
              Social and Ethical Issues of Genetic Engineering
              with Human Beings* (1982)   167

Document 79: President's Commission, *Screening and
              Counseling for Genetic Conditions* (1983)   168

Document 80: National Institutes of Health Recombinant
              DNA Advisory Committee's Subcommittee
              Statement: "Gene Therapy for Human
              Patients" (1989)   168

Document 81: Michael S. Langan, "Prohibit Unethical
              'Enhancement' Gene Therapy" (1997)   169

Document 82: Donald M. Bruce, "Moral and Ethical Issues
              in Gene Therapy" (1996)   170

Document 83: Jerome Kagan, "The Realistic View of Biology
              and Behavior" (1994)   171

Document 84: Henry T. Greely, "Genes, Patents, and
              Indigenous Peoples: Biomedical Research and
              Indigenous Peoples' Rights" (1996)   174

Document 85: A. Rural Advancement Foundation
              International, "Indigenous Person from Papua
              New Guinea Claimed in US Government
              Patent" (1995)   181

Document 86: B. Statement by Abadio Green Stocel,
              President of National Indigenous People's
              Organization of Colombia (1996)   182

Document 87: C. Statement by Ruth Liloqula, Citizen of the
              Solomon Islands (1996)   182

Document 88: David F. Betsch, "DNA Fingerprinting in
              Human Health and Society" (1994)   183

Document 89: Michelle Eadie, "Science on Trial"   183

Document 90: James L. Mudd, "Quality Control in DNA
              Typing: A Proposed Protocol" (1988)   184

Document 91: Gina Kolata, "The Code: DNA and O. J.
              Simpson: Testing Science and Justice" (1994)   185

Document 92: USA Today, "DNA Likely Was Contaminated,
              Simpson Defense Asserts" (1996)   185

Document 93: Joe Jackson, "DNA Evidence on Trial Again
              in Virginia" (1995)   186

Document 94: Robert Boyd, "DNA on File for Millions of
              Americans" (1994)   186

Document 95: "HUGO (Human Genome Organization—
              Europe) Ethics Committee, Statement on DNA
              Sampling" (1998)   187

Document 96: Mary-Claire King, "An Application of DNA
              Sequencing to a Human Rights Problem" (1991)   187

Document 97: George Annas, "Who's Afraid of the Human
              Genome?" (1989)   191

PART VI: Diagnostic Applications of Genetic Information   193

Document 98:   Richard C. Mulligan, "The Basic Science of
                Gene Therapy" (1993)   195

Document 99:   Marcia Barinaga, "Gene Therapy for Clogged
                Arteries Passes Test in Pigs" (1994)   195

Document 100:   S. Rosenberg and others, "Use of Tumor-
                Infiltrating Lymphocytes and Interleukin-2 in
                the Immunotherapy of Patients with
                Metastatic Melanoma" (1988)   196

Document 101:   Natalie Angier, "Gene Implant Therapy Is
                Backed for Children with Rare Disease" (1990)   197

Document 102:   W. French Anderson, "Human Gene
                Therapy" (1992)   197

Document 103:   Jeffrey M. Isner and others, "Arterial Gene
                Transfer of Therapeutic Angiogenesis in
                Patients with Peripheral Artery Disease" (1996) 200

Document 104:   Catherine V. Hayes, "Genetic Testing for
                Huntington's Disease—A Family Issue" (1992)   200

Document 105:   Gina Kolata, "Nightmare or the Dream of a
                New Era in Genetics?" (1993)   205

Document 106:   Gina Kolata, "Advent of Testing for Breast
                Cancer Genes Leads to Fears of Disclosure
                and Discrimination" (1997)   209

Document 107:   Kathy L. Hudson, Karen H. Rothenberg, Lori B.
                Andrews, Mary Jo Ellis Kahn, and Francis S.
                Collins, "Genetic Discrimination and Health
                Insurance: An Urgent Need for Reform" (1995)   214

Document 108:   Jeffrey M. Leiden, "Gene Therapy—Promise,
                Pitfalls and Prognosis" (1995)   221

PART VII: Ethical Issues in Genetic Engineering   225

Document 109:   Willard Gaylin, "The Frankenstein Factor"
                (1977)   227

Document 110:   James F. Keenan, "What Is Morally New in
                Genetic Manipulation?" (1990)   232

Document 111:   Thomas A. Shannon, "Ethical Issues in
                Genetic Engineering: A Survey" (1992)   234

Document 112:   John Maddox, "New Genetics Means No New
                Ethics" (1993)   240

Document 113:   Executive Summary of the NIH-DOE Working
                Group on Ethical, Legal, and Social Implications
                of Human Genome Research (1993)   243 .

Document 114:   *Diamond v. Chakrabarty* (1980)   245

Document 115:   Martin Khor, "A Worldwide Fight against
                Biopiracy and Patents on Life" (1995)   250 .

Document 116:   Mark Sagoff, "Should We Allow the Patenting
                of Life?" (1991)   251 .

Document 117:   Charles Magel, Bill of Rights of Animals
                (1998)   251

Document 118:   President's Commission for the Study of
                Ethical Problems in Medicine and Biomedical
                and Behavioral Research, *Splicing Life: The
                Social and Ethical Issues of Genetic Engineering
                with Human Beings* (1982)   252

PART VIII: Cloning   257

Document 119:   James D. Watson, "Moving toward the Clonal
                Man" (1971)   259

Document 120:   Ian Wilmut and others, "Viable Offspring
                Derived from Fetal and Adult Mammalian
                Cells" (1997)   259 .

Document 121:   Vittorio Sgaramella and Norton D. Zinder,
                "Dolly Confirmation" (1998)   260 .

Document 122:   Michael Specter with Gina Kolata, "After
                Decades and Many Mishaps, Cloning
                Success" (1997)   261 .

Document 123:   Gina Kolata, "On Cloning Humans, 'Never'
                Turns Swiftly into 'Why Not?' " (1997)   261 .

Document 124:   Gina Kolata, "In Big Advance in Cloning,
                Biologists Clone 50 Mice" (1998)   262

Document 125: Jon Cohen, "Can Cloning Help Save
               Beleaguered Species?" (1997)   263

Document 126: Kirkpatrick Sales, "Ban Cloning? Not a
               Chance" (1997)   263

Document 127: Council for Responsible Genetics, "Position
               Statement on Cloning" (1997)   264

Document 128: Pontifica Academia Pro Vita, "Reflections on
               Cloning" (1997)   264

Document 129: Constance Holden, "UN Weighs in on
               Cloning" (1997)   265

Document 130: Katharine Q. Seeyle, "Clinton Bans Federal
               Money for Efforts to Clone Humans" (1997)   266

Document 131: Munawar Ahmad Anees, "Human Cloning:
               An Atlantean Odyssey?" (1995)   266

Document 132: Rabbi Elliot Dorff, "Statement on Cloning"
               (1997)   267

Document 133: Dorothy C. Wertz, "Cloning Humans: Is It
               Ethical?" (1997)   268

Document 134: Jerome P. Kassirer and Nadia A. Rosenthal,
               "Should Human Cloning Research Be Off
               Limits?" (1998)   268

Document 135: Lori B. Andrews, "Human Cloning: Assessing
               the Ethical and Legal Quandaries" (1998)   269

Document 136: National Bioethics Advisory Commission,
               Cloning Human Beings (1997)   270

Selected Bibliography   273

Index   279

Series Foreword

This series is designed to meet the research needs of high school and college students by making available in one volume the key primary documents on a given historical event or contemporary issue. Documents include speeches and letters, congressional testimony, Supreme Court and lower court decisions, government reports, biographical accounts, position papers, statutes, and news stories.

The purpose of the series is twofold: (1) to provide substantive and background material on an event or issue through the texts of pivotal primary documents that shaped policy or law, raised controversy, or influenced the course of events, and (2) to trace the controversial aspects of the event or issue through documents that represent a variety of viewpoints. Documents for each volume have been selected by a recognized specialist in that subject with the advice of a board of other subject specialists, school librarians, and teachers.

To place the subject in historical perspective, the volume editor has prepared an introductory overview and a chronology of events. Documents are organized either chronologically or topically. The documents are full text or, if unusually long, have been excerpted by the volume editor. To facilitate understanding, each document is accompanied by an explanatory introduction. Suggestions for further reading follow the document or the chapter.

It is the hope of Greenwood Press that this series will enable students and other readers to use primary documents more easily in their research, to exercise critical thinking skills by examining the key documents in American history and public policy, and to critique the variety of viewpoints represented by this selection of documents.

Significant Dates in the History of Genetics

1672   Reinier de Graff describes the human egg.

1759   Caspar Wolff discovers the function of the human egg.

1767   Antoni Van Leeuwenhook describes the human sperm.

1865   Gregor Mendel conducts his studies on dominant and recessive traits in peas and publishes his results, which go unnoticed.

1882   Prof. August Weismann of Germany discovers the equal contribution of male and female in reproduction and the role of the chromosomes.

1895   Mendel's contributions are finally recognized and begin to influence studies of inheritance.

1896   Prof. E. Wilson of the United States develops a theory of inheritance based specifically on the role of chromosomes.

1902   Dr. Archibald Garrod of England demonstrates the inheritance of disorders through several generations of the same family.

1903   German and U.S. scientists independently validate that units of inheritance must be on the chromosome.

1908   Scientists discover that genes can act on other genes and that their effects can be cumulative.

1910   Prof. Thomas Morgan of Columbia University in New York shows that some traits travel on particular chromosomes, the first demonstration of sex-linked characteristics.

1933   Prof. T. S. Painter learned how to dye chromosomes and showed that each has its distinct pattern and that they can be distinguished from each other.

1953   James Watson and Francis Crick, working in England, discover the famous double helix structure of DNA, thus revolutionizing modern biology.

1970   Prof. Stanley Cohen of Stanford University discovers how to cut and recombine strands of DNA from different organisms, thus creating the technology of recombinant DNA.

1973   First conference to discuss the idea of gene mapping to determine the site of specific chromosomes.

1974   The first Asilomar conference called by scientists to discuss the social implications of their work, which led to a one-year moratorium on research to allow the development of safety guidelines.

1976   Guidelines for the conducting of research using the recombinant DNA technology are published in the United States by the National Institutes of Health, which also established the Recombinant DNA Molecule Program Advisory Committee.

1980   In *Diamond v. Chakrabarty*, 447 U.S. 303, the U.S. Supreme Court approves the first patenting of a genetically engineered organism, an oil eating bacterium developed by General Electric.

1984   Prof. Mario Capecchi of the University of Utah develops "knock-out mouse" technology, which allows the production of a mouse that is missing a particular gene.

1985   Prof. Kerry Mullins discovers the polymerease chain reaction, which allows the rapid multiplication of sections of DNA to obtain large samples for study.

1988   U.S. Patent Office approves a patent on a genetically engineered mouse developed at Harvard University to be used for cancer studies.

1988   The funding of the Human Genome Project by the U.S. Congress to map the entire genome. James Watson is named the first director.

1990   Approval in the United States for the first use of gene therapy in which engineered genes are inserted into the blood to cure disease.

1997   The announcement of the first cloning of a mammal, the sheep Dolly, by Dr. Ian Wilmut of the Roslin Institute, Scotland.

1998   Dr. Ryuzo Yanagimachi of the University of Hawaii announces a technique through which he cloned 50 mice, some of which are clones of the clones.

Introduction

The progress made in the field of genetic engineering over the last decade, even the last five years, makes us forget how long it took to develop the first critical insight into the units of inheritance that we now call genes. Once those key insights were made, then the knowledge and technology explosion that followed was almost inevitable. Examining the following outline of the main developments in the history of genetics will situate the reader to the main focus of this volume: to examine, from a variety of perspectives, current developments in genetic engineering, particularly the interventions that can replace the genes of one species with those of another, the breeding of organisms with or without particular genes, and the genetic modification of an organism with a gene from another species so that its descendants inherit that gene and in turn pass it on to its descendants. These developments required a fundamental insight, and it is to the history of this insight and its consequences that we now turn.

People know that historically there were such things as family resemblances in the plant, animal, and human world. Selective breeding occurred in the animal world and even in the human world. We read, for example, in Plato's *Republic* that the rulers or guardians should be selected from the best citizens and then selectively bred to ensure the continuity of quality leadership. Later thinkers, under Plato's influence, developed the theory that an individual developed through a vital force transmitted in the male's semen that contained the form of the human. Only in 1600 with the invention of the microscope was the British physician William Harvey, who also discovered the circulation of blood in the body, able to discover eggs as well as sperm. Refinements of the microscope allowed Caspar Wolff in the mid-1700s to make careful observations of the chicken embryo and the development of a fertilized egg.

As important as these discoveries of the existence of sperm and egg and their role in reproduction were, they did not begin to answer the important questions of how and which characteristics were inherited. (Although Plato's eugenic theory appalled many, he did nevertheless point to a particular problem that continues to our day: What, exactly, can be inherited? Can our personality traits be inherited? Can wisdom, generosity, cunning, shrewdness, or tact be inherited? Or is it only our physical characteristics such as eye color, height, physical profile, or tendency to baldness that is inherited? While this question continues to be debated intensely, one answer about inheritance has been resolved.)

(The question about the inheritance of physical characteristics was answered by the work of the German Augustinian monk Gregor Mendel (1822–1884).) Mendel engaged in a series of experiments with peas he grew in the monastery garden. These experiments demonstrated that units of heredity were physically passed from one generation to the next. He did this by tracing various characteristics through successive generations. But just as important, he discovered that a particular trait could sometimes not show up for several generations and then reappear in successive generations. These are the familiar concepts of dominant and recessive traits. From this Mendel concluded "that each parent plant contributed one of a pair of factors, and no more than one, to determining each character in the makeup of each offspring plant."[1] Thus was the critical idea of modern genetics born in a monastery. Mendel's results were published in 1865 but went unnoticed for some 30 years.

As noted above, other developments occurred: the discovery of sperm and eggs and their roles, the fusing of eggs and sperm in fertilization, and the discovery in 1877 of small structures in the nucleus of the cell as it is to divide. These structures were called chromosomes. Like Mendel's peas, each new individual receives its full set of chromosomes, one half from the father and one half from the mother. Finally, in 1882 the German scientist August Weismann wrote the book *Das Keimplasma* (germplasm), in which he showed that "the male and the female parent contribute equally to the heredity of the offspring," that sexual reproduction thus generates a new combination of hereditary factors, and that the chromosomes must be the bearers of heredity.[2] By 1896 E. Wilson, an American biologist, developed a theory of inheritance based on the role of the chromosomes.

In 1900 Mendel's paper was discovered and his ideas given exposure to the scientific community. In 1902 a British physician, Archibald Garrod, reported a disorder he had been able to trace through several generations. This was the first demonstration of a characteristic to be inherited in humans through the use of Mendelian principles.[3] Diseases are now known to be inherited in this fashion. Then in 1903 German and American scientists independently validated that the units of inheritance must be on the chromosomes themselves. In 1908 other scientists showed that the actions of "some genes modify the actions of other genes."[4] That is, whereas some genes act independently, others work in groups and their effect is cumulative. Then in the lab of Thomas Morgan at Columbia, the discovery was made, in 1910, that some traits travel on particular chromosomes. Morgan showed that eye color in fruit flies is sex-linked. In humans we have the example of red-green color-blindness or hemophilia, as well as other diseases that travel on the X chromosome. And in 1933 T. S. Painter used a dyeing technique to show that each pair of chromosomes has its own distinct pattern and that every pair can be distinguished from the other 46 pairs.

All this work laid the foundation for much exciting and detailed work in the biology of inheritance and established many critical directions in genetic research. The core elements were known, and attention now focused on the biochemistry of the action of the genes to understand their function better. There remained many critical questions, however, among them the question of how genetic information replicates itself. Attention had been called to deoxyribonucleic acid, or DNA, as a possible answer, but no proof was found.

But in the 1950s an American, James Watson, and a British colleague, Francis Crick, working in the Cavendish Laboratory in Cambridge, discovered the now famous double helix structure of the DNA molecule which in many ways proved to be the Rosetta Stone of modern genetics. The story of their discovery, told in Watson's book *The Double Helix*, reads like both a mystery story and a comedy of errors and provides a rare and frank vision of the competitive world of very high level research science. Their discovery was reported in a two-column-long letter in the British journal *Nature*, in which they laconically noted in the last sentence that the predictive power of this structure for inheritance had not escaped their notice.

What Watson and Crick explained with their model was how the DNA molecule replicates itself and how it directs the structure of proteins that form the basis of further biological developments. They found that the structure of the DNA molecule consists of the bases adenine and thymine and cytosine and guanine, which always pair with each other to provide the sequence in which hereditary information is encoded and passed on to one's descendants. This led to further discoveries of how DNA replicates itself and how to read the sequences of the bases in the molecule and ultimately to the project to map the location of the hundreds of thousands of base pairs in the human genome.

This discovery of the double helix structure of the DNA molecule opened the flood gates of research and discovery, and the next decade

witnessed the discovery of the fundamental explanations of the genetically directed process of the cell. What this led to in the 1970s was the
development of a variety of techniques to gain access to the gene and to
manipulate its contents. Of particular importance was learning how the
genes switch themselves off and on so that one cell becomes a liver,
another an eye, another the stomach, and yet another the brain. First the
biologists learned how to make multiple copies of genes through the
introduction of the gene into bacteria; "the bacteria then can multiply
normally with the foreign DNA multiplying in step."[5] This is called cloning, the making of many identical copies of a single cell. Then scientists
learned that bacteria have the capacity to disable foreign DNA. This capacity is lodged in what are referred to as restriction enzymes, so called
because "they restrict the range of hosts in which a virus can survive."[6]
What is more interesting is that these restriction enzymes cut DNA at
particular places and thus act as chemical scissors to cut DNA so that
another segment can be inserted into this segment. This process is called
recombinant DNA because two types of DNA are recombined with each
other.

This discovery opened the door to the possibility of genuine genetic
engineering: to transfer genes from one species to another and to construct a variety of new organisms not present in nature. The positive
benefits of this research were recognized, but so were its harmful effects,
particularly for intentionally or unintentionally creating pathogens or
other organisms harmful to humans or the environment. Concerns about
the harmful applications of this new technology gave rise to a conference
at Asilomar, California, in 1974 at which scientists, ethicists, and attor-
neys discussed both the science and the social implications of the technology of recombinant DNA. That this conference was even held to
discuss these concerns was significant enough. But what emerged from
it was astounding. The scientists agreed to a voluntary moratorium on
research for a year during which safety guidelines would be established,
the technologies studied to determine their risks, and a variety of safety
procedures for the laboratories developed. In addition, scientists developed new biological material to use in the study of gene function. One
such organism was engineered to self-destruct if exposed to sunlight.
Thus even if an experimental and potentially dangerous organism escaped the lab, it would not cause harm because it could self-destruct
outside the lab setting.

The National Institutes of Health (NIH), in conjunction with many
scientists working in this field, developed a set of guidelines for recombinant DNA technology. As with all guidelines of this type, not all were
satisfied. Some scientists found them too restrictive, some found them
appropriate, and some wished that the Asilomar conference had never
been held at all. Several of these concerns, as well as a discussion of the

technologies and the guidelines, can be found in the article by Stanley Cohen, one of the discoverers of the recombinant DNA technology, in Part I of this reader. One main point of tension in the safety discussions was that many of the risks discussed were hypothetical or set in a worst-possible-case scenario. Such scenarios do not always permit the calmest of discussions, nor do they help a poorly or moderately informed public focus on the appropriate issues. The city of Cambridge, Massachusetts, under the direction of its mayor and city council, held a series of public hearings more noted for heat than light, but which eventually resulted in legislation that imposed exceptionally strict standards on recombinant DNA research done at Harvard and MIT. It should be noted that there has never been a reported safety problem or biohazard reported with recombinant DNA research in the two decades of international research. While the concerns for safety discussed at the beginning of the applications of these technologies were quite appropriate because of the newness of the technologies and an absolute lack of experience with them, the ensuing history of research in genetics has been conducted responsibly and with an excellent safety record.

Within this debate the attention has now shifted from the focus on the design of organisms and the consequences of accidental releases of organisms from the labs to the deliberate introduction of genetically engineered organisms, such as various grains, into the environment or the using of genetically engineered plants and animals as food products for humans and animals. From the agricultural perspective, the questions are standard ones: Will such a hybrid plant be successful in a normal environment? Will it drive out other naturally occurring varieties of the same plant? Could the genetically altered plant breed with other naturally occurring varieties and ultimately drive out the desired plants? The purpose of developing genetically altered plants, especially grains, is to produce grains with higher yields, grains that are disease or frost resistant and can survive in new environments. Controlled experiments have been done with restrictive plantings in outdoor fields, but many of the questions cannot be answered until widespread plantings have been done and the outcomes monitored carefully. The issue here is to do such plantings in as carefully a controlled way as possible and to monitor the results. Those developing the grains do not see major risks or safety concerns for the environment or for the status of other grains. The experiments thus far have given no justification for major concern; but as with the development of recombinant DNA techniques, the field is new and care is warranted.

The same is true with the introduction of genetically engineered foods or food products into the human and animal food supply. The core question is whether such products will harm animals, which may in turn cause harm to humans when they are eaten, or harm humans directly.

Tomatoes genetically engineered to be frost-resistant and to have delayed ripening are on the market. Some people have rejected them because they are not natural products; others resist them because they claim the products are unsafe. In both the United States and Europe there are large campaigns against such products. Although no instances of actual harm from the use of such products has been documented, a case that indicates the need for vigilance did arise. Soy products are a major human and animal food source. The soybean, however, lacks a certain amino acid beneficial to animals. Thus when animals are fed soy products, they also need to be given supplements. A company took the gene for this particular amino acid from the Brazil nut and spliced it into the soybean so that the bean now contained this nutrient and thus supplements would no longer be necessary. However, flour for human consumption would also be made from these soybeans, and Brazil nuts are a known allergin. The question was, would the allergenic properties of the Brazil nut be transferred to the soybean? When humans were exposed to the genetically altered soybean, allergenic reactions did indeed occur, and the product was withdrawn from the market as a result. The example demonstrates two things: Care needs to be exercised in developing genetically engineered foods, and industry can be responsive to safety concerns. These and other questions are examined in Part III of this reader. Thus far the record is good, but problems remain and call for continued vigilance.

The next major step was the development of a map of the human genes. A critical technical development that opened the door to this was a technique discovered by Mary Weiss and Howard Green called somatic-cell hybridization. They discovered that if one adds a particular mammalian-tumor virus, the Sendai virus, to a mixture of human and mouse cells, the cell walls weaken and some cells fuse and multiply. This technique allows the multiplication of human chromosomes and the creation of different cell lines with different fragments of human chromosomes. This access to many chromosomes made locating various genes possible.[7] This then opened the way to other technical developments in mapping technologies, and the first human gene mapping conference was held in 1973. Genes were located on specific chromosomes and subsections of them. This led to the next phase: the study of the anatomical makeup of the gene itself at the level of the various molecules of the DNA. This was based on the variations in DNA called restriction fragment length polymorphisms, more commonly referred to as RFLPS. Then scientists learned that "when a restriction enzyme is applied to DNA from different individuals, the resulting sets of fragments sometimes differ markedly from one person to the next."[8] This means that mutations in the sequence of the bases vary from person to person. This also means that "variations anywhere in the neighborhood of a gene could usefully serve as markers."[9] Once markers were located, the particular gene responsible for the variations could be located. The result of this has been the discovery of the single gene responsible for a variety of diseases such as Huntington's disease, Duchenne muscular dystrophy, and most recently some forms of breast cancer.

The next phase of mapping required the development of techniques to handle large batches of DNA, to replicate DNA sequences, and to do this more efficiently. Indeed, the first estimates of the cost of mapping the genome were exceptionally high because the work had to be done manually by various lab technicians. But new discoveries for rapid and accurate replications of DNA sequences outside of a cell, called polymerease chain reaction (PCR), were made. To give an example of the speed of replication, the scientist begins with a specially prepared sample of DNA; after twenty cycles of treatment, "the portion of DNA including the target has been multiplied a million-fold."[10] This process was developed by Kerry Mullins, who worked at Cetus Corporation, which patented the process and then later sold it to Hoffman-LaRouche, Inc., for $300 million. One can see that not only are scientific reputations to be made, but so are personal and corporate fortunes.

(The current state of the art is the application of these and many more techniques to continue the process of mapping the human genome.] The process is complex and time consuming. "The DNA sequence has a simple numerical expression: it is composed of three billion base pairs. That is enough information to code for about 100,000 to 300,000 genes, each gene being a region of DNA that can specify a protein or some other structure that carries out a function in the organism."[11] What this map will give us is a picture of the essential elements that specifies us as humans, showing what makes us work and what differentiates us from other animals. It will also allow us to begin to discover what particular genes do, what various combinations of genes do, how various environments can affect the expression of genes, and, of critical importance for many people, how to intervene to cure disease at the genetic level.

(The work on this project is not easy even though the effort is international and vast sums of funds have been allocated. In the United States, for example, about $15 billion has been allocated for a ten-year period.)(Even more interesting, in the United States about 3 percent of that budget has been allocated to investigate ethical, legal, and social issues raised by this research. Thus the Human Genome Project in the United States is focused not only on scientific discoveries but also on the implications of those same scientific discoveries.)

(The 1997 publication of the first successful cloning of a mammal by Dr. Ian Wilmut of the Roslin Institute in Scotland added to the scientific and ethical debates over genetics.)The primary fear was that this technology would be applied to humans, which raises many ethical problems.(This debate was further enhanced by the 1998 publication by Dr. Ryuzo Yanagimachi of the University of Hawaii of a technology that facilitated the cloning of 50 mice) some of which were clones of the clones. This shows that the work of Wilmut was not a fluke, and scientists expect more rapid developments in cloning technology, raising yet more concerns about the application to humans.

That concern, and many others noted in this introduction, is reflected in many of the documents in the various sections of this reader. Whereas some articles set out the basic science and technology and give descriptions of various applications of these discoveries, other articles look at ethical, legal, and social issues raised by these developments. There are raised, for example, questions about the control of genetic information about one's self, the implications of intervening to change one's genetic profile even for therapeutic purposes, the development of genetically altered animals for research and genetically altered grains and vegetables for human consumption, and the patenting of various discoveries. Many of these questions are challenging both scientifically and ethically. Yet they raise some of the most important and controversial issues of our day. The problems the new genetics raises are not esoteric problems for scientists; they are problems most of us will have to face. Will we all be genetically screened to determine our future health profile? Can we be denied insurance if it is discovered that we are disposed to certain diseases, even though the diseases may not develop? Will we utilize the latest prenatal genetic tests to screen pregnancies for diseases as well as for the sex of the fetus? Will the knowledge that we have a late-onset genetic disease be helpful or harmful to us?

The map of our genes will tell us much about who we are and how we function, but how will we react to this knowledge and how will we use it? These are the critical questions with which informed citizens will want to come to terms.

NOTES

1. Horace F. Judson, "A History of the Science and Technology Behind Gene Mapping and Sequencing." In Daniel J. Kevles and LeRoy Hood, eds., *The Code of Codes: Scientific and Social Issues in the Human Genome Project* (Cambridge, MA: Harvard University Press, 1992), p. 41.

2. Judson, "A History of Gene Mapping," p. 42.
3. Some 5600 genes.
4. Judson, "A History of Gene Mapping," p. 43.
5. Judson, "A History of Gene Mapping," p. 62.
6. Judson, "A History of Gene Mapping," p. 63.
7. Judson, "A History of Gene Mapping," pp. 68–69.
8. Judson, "A History of Gene Mapping," p. 70.
9. Judson, "A History of Gene Mapping," p. 70.

10. Judson, "A History of Gene Mapping," p. 76.

11. Walter Gilbert, "A Vision of the Grail." In Daniel J. Kevles and LeRoy Hood, eds., *The Code of Codes: Scientific and Social Issues in the Human Genome Project* (Cambridge, MA: Harvard University Press, 1992), p. 83.

Part I

The Debates Surrounding Genetic Engineering

The following documents provide several different aspects of the debates that accompanied the beginnings of genetic engineering. The center of attention is the process called recombinant DNA (rDNA) research. As will be described in several articles, this is a process by which genetic material is taken from one organism and placed into another. Of singular importance in this technology is that the genetic material spliced into an organism need not be from the same organism or even from the same species as the organism. Thus we have the possibility of the production of genuinely new organisms.

Whereas the benefits of this research, such as the development of new drugs and vaccines, were apparent to many individuals, others were concerned about such an engineered organism's escaping into the environment and causing harm. The reason for such concern was that such a situation had never been experienced before. These concerns gave rise to many debates and eventually led to the development of regulations as well as an oversight committee in the National Institutes of Health (NIH).

The documents present the ideas of scientists working in this area as well as the reaction to them and examples of regulations governing this type of research.

DOCUMENT 1: James Watson, *The Double Helix* (1968)

This document presents a selection from this scientific autobiography in which Watson describes the events that led to the discovery of the structure of the DNA molecule.

* * *

When I got to our still empty office the following morning, I quickly cleared away the papers from my desk top so that I would have a large, flat surface on which to form pairs of bases held together by hydrogen bonds. Suddenly I became aware that an adenine-thymine pair held together by two hydrogen bonds was identical in shape to a guanine-cytosine pair held together by at least two hydrogen bonds. All the hydrogen bonds seemed to form naturally; no fudging was required to make the two types of base pairs identical in shape.

Even more exciting, this type of double helix suggested a replication scheme much more satisfactory than my briefly considered like-with-like pairing. Given the base sequence on one chain, that of its partner was automatically determined. Conceptually, it was thus very easy to visualize how a single chain could be the template for the synthesis for a chain with the complementary sequence.

However, we both knew that we would not be home until a complete model was built in which all the stereo-chemical contacts were satisfactory. There was also the obvious fact that the implications of its existence were far too important to risk crying wolf. Thus I felt slightly queasy when at lunch Francis winged into the Eagle to tell everyone within hearing distance that we had found the secret of life.

Source: James Watson, *The Double Helix* (New York: Atheneum, 1968).

DOCUMENT 2: J. D. Watson and F. H. Crick, "Molecular Structure of Nucleic Acids" (1953)

This document is an excerpt from the two-column letter to the editor in which Watson and Crick announced the now famous double helix structure of the DNA molecule. Of particular interest is the brevity of the article; it was a letter to the editor and won Watson and Crick the Nobel Prize.

* * *

We wish to suggest a structure for the salt of deoxyribose nucleic acid (D.N.A.). This structure has novel features which are of considerable biological interest.

We wish to put forward a radically different structure for the salt of deoxyribose nucleic acid. This structure has two helical chains each coiled round the same axis.

It is assumed that the bases only occur in the structure in the most plausible tautomoric forms. . . . These pairs are: adenine (purine) with thymine (pyrimidine), and guanine (purine) with cytosine (pyrimidine).

In other words, if an adenine forms one member of a pair, on either chain, then on these assumptions, the other member must be thymine; similarly for guanine and cytosine. The sequence of bases on a single chain does not appear to be restricted in any way. However, if only specific pairs of bases can be formed, it follows that if the sequence of bases is given, then the sequence on the other chain is automatically determined.

It has not escaped our notice that the specific pairing we have postulated immediately suggests a possible copying mechanism for the genetic material.

Source: J. D. Watson and F. H. Crick, "Molecular Structure of Nucleic Acids," *Nature* 4356 (April 25, 1953): 737.

DOCUMENT 3: Robert Sinsheimer, "Troubled Dawn for Genetic Engineering" (1975)

In his article "Troubled Dawn for Genetic Engineering," Robert Sinsheimer explores three advances in the new field of DNA technology and some of their applications. The following excerpt from this article presents questions he and others raised that gave rise to 1975 Asilomar conference, held in California, which established a voluntary moratorium on some DNA research. The article concludes with a consideration of the implications of intervening in the human gene pool.

* * *

The essence of engineering is design and thus, the essence of genetic engineering, as distinct from applied genetics, is the introduction of human design into the formulation of new genes and new genetic combinations. These methods thus supplement the older methods which rely upon the intelligent selection and perpetuation of those chance genetic combinations which arise in the natural breeding process.

The possibility of genetic engineering derives from major advances in DNA technology in the means of synthesizing, analyzing, transposing and generally manipulating the basic genetic substance of life. Three major advances have all neatly combined to permit this striking accomplishment: these are, 1, the discovery of means for the cleavage of DNA at highly specific sites; 2, the development of simple and generally applicable methods for the joining of DNA molecules; and 3, the discovery of effective techniques for the introduction of DNA into previously refractory organisms.

It is very probable that in time the appropriate genes can be introduced into bacteria to convert them into biochemical factories for producing complex substances of medical importance: for example, insulin (for which a shortage seems imminent), growth hormone, specific antibodies, and clotting factor VIII, which is defective in hemophiliacs. Even if these specific genes cannot be isolated from the appropriate organisms, the chances of synthesizing them from scratch are now significant.

Other more grandiose applications of microbial genetic engineering can be envisaged. The transfer of genes for nitrogen fixation into presently inept species might have very significant agricultural applications. Appropriate design might permit appreciable modifications of the normal bacterial flora of the human mouth with a significant impact upon the incidence of dental caries. Even major industrial processes might be carried out by appropriately planned microorganisms.

However, we must remember that we are creating here novel, self-propagating organisms. And with that reminder, another darker side appears on this scene of brilliant scientific enterprise. For instance, for scientific purposes there is great interest in the insertion of particular regions of viral DNA into plasmids—particularly, portions of oncogenic (cancer-inducing) viral DNA—so as to be able to obtain such portions and their gene products in quantity and subsequently to study the effects of these substances on their normal host cells. Abruptly we come to the potential hazard of research in this field, in fact the specific hazard which inspired the widely known "moratorium" proposed last year by a committee of the US National Academy, chaired by Paul Berg.

This moratorium and its related issues deserve very considerable discussion. Briefly, it became apparent to the scientists involved—at almost the last hour when all of the techniques were really at hand—that they were about to create novel forms of self-propagating organisms—derivatives of strains known to be normal components of the human intestinal flora—with almost completely unknown potential for biological havoc. Could an Escherichia coli strain carrying all or part of an oncogenic virus become resident in the human intestine? Could it thereby become a possible source of malignancy? Could such a strain spread throughout a human population? What would be the consequence if even an insulin-secreting strain became an intestinal resident? Not to mention the more malign or just plain stupid scenarios such as those which depict the insertion of the gene for botulinus toxin into Escherichia coli.

Unknown probabilities

Unfortunately the answers to these questions in terms of probabilities that some of these strains could persist in the intestines, the probabilities that the modified plasmids might be transferred to other strains better adapted to intestinal life, the probabilities that the genome of an oncogenic virus could escape, could be taken up, could transform a host cell, are all largely unknown.

Following the call for a moratorium a conference was held at Asilomar at the end of last February to assess these problems. While it proved possible to rank various types of proposed experiments with respect to potential hazard, for the reasons already stated it proved impossible to establish, on any secure basis, the absolute magnitude of hazard. Various distinguished scientists differed very widely, but sincerely, in their estimates. Historical experience indicated that simple reliance upon the physical containment of these new organisms could not be completely effective.

In the end a broad, but not universal, consensus was reached which recommended that the seemingly more dangerous experiments deferred until means of "biological containment" could be developed to supplement physical containment. By biological containment is meant the crippling of all vehicles—cells or viruses—intended to carry the recombinant genomes through the insertion of a variety of genetic defects so as to reduce very greatly the likelihood that the organisms could survive outside of a protective, carefully supplemented laboratory culture.

Nor was there any sustained discussion at Asilomar of ancillary issues such as the absolute right of free inquiry claimed quite vigorously by some of the participants. Here, I think, we have come to recognize that there are limits to the practice of any human activity. To impose any limit upon freedom of inquiry is especially bitter for the scientist whose life is one of inquiry: but science has become too potent. It is no longer enough to wave the flag of Galileo.

Rights are not found in nature. Rights are conferred within a human society and for each there is expected a corresponding responsibility. Inevitably at some boundaries different rights come into conflict and the exercise of a right should not destroy the society that conferred it. We recognize this in other fields. Freedom of the press is a right but it is

subject to restraints, such as libel and obscenity and, perhaps more du-
biously, national security. The right to experiment on human beings is
obviously constrained. Similarly, would we wish to claim the right of
individual scientists to be free to create novel self-perpetuating organisms likely to spread about the planet in an uncontrollable manner for
better or worse? I think not.

This does not mean we cannot advance our science or that we must
doubt its ultimate beneficence. It simply means that we must be able to
look at what we do in a mature way. There was, at Asilomar, no explicit
consideration of the potential broader social or ethical implications of
initiating this line of research—of its role, as a possible prelude to longer-
range, broader-scale genetic engineering of the flora and fauna of the
planet, including, ultimately, man. It is not yet clear how these techniques may be applied to higher organisms but we should not under-
estimate scientific ingenuity. Indeed the oncogenic viruses may provide
a key; and mitochondria may serve as analogues for plasmids.

Controlled evolution?

How far will we want to develop genetic engineering? Do we want to
assume the basic responsibility for life on this planet—to develop new
living forms for our own purpose? Shall we take into our own hands
our own future evolution? These are profound issues which involve science but also transcend science. They deserve our most serious and continuing thought. I can here mention only a very few of the more salient
considerations.

Clearly the advent of genetic engineering, even merely in the microbial
world brings new responsibilities to accompany the new potentials. It is
always thus when we introduce the element of human design. The distant yet much discussed application of genetic engineering to mankind
would place this equation at the centre of all future human history. It
would in the end make human design responsible for human nature. It
is a responsibility to give pause, especially if one recognizes that the
prerequisite for responsibility is the ability to forecast, to make reliable
estimates of the consequence.

Can we really forecast the consequence for mankind, for human society, of any major change in the human gene pool? The more I have
reflected on this the more I have come to doubt it. I do not refer here to
the alleviation of individual genetic defects—or, if you will, to the oc-
casional introduction of a genetic clone—but more broadly to the genetic
redefinition of man. Our social structures have evolved so as to be more
or less well adapted to the array of talents and personalities emergent
by chance from the existing gene pool and developed through our cultural agencies. In our social endeavors we have, biologically, remained
cradled in that web of evolutionary nature which bore us and which has

undoubtedly provided a most valuable safety net as we have in our fumbling way created and tried out varied cultural forms.

To introduce a sudden major discontinuity in the human gene pool might well create a major mismatch between our social order and our individual capacities. Even a minor perturbation such as a marked change in the sex ratio from its present near equality could shake our social structures—or consider the impact of a major change in the human lifespan. Can we really predict the results of such a perturbation? And if we cannot foresee the consequence, do we go ahead?

It is difficult for a scientist to conceive that there are certain matters best left unknown, at least for a time. But science is the major organ of inquiry for a society—and perhaps a society, like an organism, must follow a developmental programme in which the genetic information is revealed in an orderly sequence.

The dawn of genetic engineering is troubled. In part this is the spirit of the time—the very idea of progress through science is in question. People seriously wonder if through our cleverness we may not blunder into worse dilemmas than we seek to solve. They are concerned not only for the vagrant lethal virus or the escaped mutant deadly microbe, but also for the awful potential that we might inadvertently so arm the anarchist in our society as to shatter its bonds or conversely so arm the tyrannical in our society as to forever imprison liberty.

It is grievous that the elan of science must be tempered, that the glowing conviction that knowledge is good and that man can with knowledge lift himself out of hapless impotence must now be shaded with doubt and caution. But in this we join a long tradition. The fetters that are part of the human condition are not so easily struck.

We confront again, the enduring paradox of emergence. We are each a unit, each alone. Yet, bonded together, we are so much more. As individuals men will have always to accept their genetic constraints, but as a species we can transcend our inheritance and mold it to our purpose—if we can trust ourselves with such powers. As geneticists we can continue to evolve possibilities and take the long view.

Source: Robert Sinsheimer, "Troubled Dawn for Genetic Engineering," *New Scientist*, October 1975, 148–51.

DOCUMENT 4: Paul Berg, D. Baltimore, S. Brenner, R. Roblin, and M. Singer, "Asilomar Conference on Recombinant DNA Molecules" (1975)

This document reports findings on safety issues considered by the second conference at Asilomar, California, in 1975. While recognizing that some potential hazards remained unknown, scientists had obtained enough experience to propose conditions under which research could go forward. These recommendations also formed the basis of future NIH guidelines.

* * *

Nevertheless, the participants at the Conference agreed that most of the work on construction of recombinant DNA molecules should proceed, provided that appropriate safeguards, principally biological and physical barriers adequate to contain the newly created organisms, are employed. Moreover, the standards of protection should be greater at the beginning and modified as improvements in the methodology occur and assessments of the risks change.

[The document then identifies four levels of risk with the type of containment appropriate to each.]

1. Minimal risk

Essential feature of such facilities are not drinking, eating, or smoking in the laboratory, wearing laboratory coats in the work area. . . .

2. Low risk

The key features of this containment are a prohibition of mouth pipetting, access limited to laboratory personnel, and the use of biological safety cabinets. . . .

3. Moderate risk

The principal features of this level of containment are that transfer operations should be carried out in biological safety cabinets, gloves should be worn during the handling of infectious materials, vacuum lines must be protected by filters, and negative pressure should be maintained in the limited access laboratories.

4. High risk

The main features of this type of facility are its isolation from other areas by air locks, a negative pressure environment, a requirement for clothing changes and showers for entering personnel, and laboratories fitted with treatment systems to inactivate or remove biological agents.

Source: Paul Berg, D. Baltimore, S. Brenner, R. Roblin, and M. Singer, "Asilomar Conference on Recombinant DNA Molecules," *Science* 188 (June 6, 1975): 991–994.

DOCUMENT 5: Daniel Callahan, "Recombinant DNA: Science and the Public" (1977)

Because of concern over potential personal and environmental harms from rDNA research, the National Institutes of Health developed guidelines under which such research would be conducted. The following document presents the background leading to the development of the first guidelines related to DNA research, particularly the public dimension of this debate. The author identifies three questions relevant to this debate: Why did the public take so long to become involved? What options are open to the public and how might they be evaluated? What ethical and social criteria might be used to evaluate such research?

* * *

What has been the relationship between science and the public on the issue of a recombinant DNA?

Although there were earlier informal events, I will date the beginning as the Gordon Conference, a scientific meeting on nucleic acids held in New England, in the summer of 1973. As a result of concerns expressed by a group of scientists at that conference, an open letter was sent to *Science*, requesting the National Academy of Sciences to establish a committee to study various problems of recombinant DNA research and to recommend specific actions or guidelines in the light of potential hazards. That was and remains a striking act of moral initiative.

A committee was formed, which recommended that a moratorium on certain forms of recombinant DNA research be established voluntarily. It also recommended that the National Institutes of Health (NIH) set up an advisory committee to evaluate potential hazards in the research, to devise safety procedures, to develop guidelines for researchers working with potentially hazardous DNA molecules, and to call an international conference. In October 1974 the NIH established the Recombinant DNA Molecule Program Advisory Committee. In February 1975, the international conference was held at the Asilomar Conference Center in Pacific Grove, California. In essence, the conference participants concluded that the voluntary moratorium should be lifted and that future research should be conducted under a set of guidelines. Immediately thereafter, the NIH Advisory Committee began work on refining the guidelines suggested at Asilomar.

Where was the public up to this point? Not much in evidence. Save for the presence of four lawyers at Asilomar and the presence of a good number of reporters, the public was at that point little involved. What are we to make of that fact? Nothing very portentous, and certainly nothing deceptive. The very first thrust of the recombinant DNA debate was within the scientific community. The group now called Berg et al. wanted to raise a moral issue in the scientific community: some of the proposed recombinant DNA research could be dangerous. They succeeded in putting that moral issue on the table, and not without opposition. They recognized the significance for the public of what was, at that stage, a struggle among scientists trying to determine whether or not there were real dangers in the research. They signaled that quite clearly by inviting the lawyers and the press to Asilomar.

Surely, one might say, a handful of lawyers and reporters at a closed event hardly constitutes a full involvement of the public. True enough, but that is all retrospective wisdom. In the first place, the whole problem was new. There exist no real historical precedents, including the development of the atom bomb, for a problem of this kind: what ought one to do in a situation where spectacular hypothetical possibilities are balanced against equally spectacular hypothetical dangers? In the second place, there are no ready-made forums for public discussions of such matters. The scientists had to create their own. Had they not done so, and there would probably never have been debates in Cambridge, Ann Arbor, and elsewhere, and we would probably not be here today.

NIH Guidelines

During 1975 the NIH Advisory Committee developed guidelines, and in December sent a draft to the Director of NIH, Dr. Donald Fredrickson. There were no representatives of the public on the NIH Advisory Committee, though a lay person was added later. That was a real oversight. Moreover, the committee was essentially studying technical, not policy matters. The next important event was the meeting of the Director's Advisory Committee on February 9–10, 1976. To this meeting were invited representatives of various public interest groups, representatives of various factions within the scientific community, and other interested parties. At the meeting the draft guidelines were examined publicly. On the whole, it was a subdued debate.

However low-keyed that meeting at NIH, the ingredients of a coming storm were present. There was, for one thing, a disturbing realization on the part of some that the debate and discussion on guidelines had moved very fast. Why was it, for instance, that the first fully public discussion of recombinant DNA was already focused on draft guidelines? A decision even to prepare and discuss guidelines seemed to assume that the basic ethical, political and social questions about recombinant DNA research had all been raised and fully discussed. But they had not been.

For another thing, while there were certainly public representatives at the NIH meeting, and many reporters as well, it was hardly likely in such a genteel setting that the full voice of the public could have been heard, or that, if heard, it could have significantly slowed down the rapidly moving machinery of putting guidelines in place to govern NIH grants. It should hardly be astounding that some felt vested interests were pulling the strings and that, once again, the scientific establishment was manipulating the public.

I reject that cynical interpretation. NIH's immediate problem was to set some ground rules on a form of scientific research which was moving very rapidly and which, one way or the other, had to be controlled and monitored. The guidelines represent a compromise solution. But, some have complained, in such a potentially fateful situation, compromise is not good enough. They may be right—future generations will make that judgment—but the reality now is that neither the public nor the scientific community share any consensus whatever on recombinant DNA research. And what do we do in our society when there is no consensus on an issue? We normally compromise, for that is the one way we can continue to live together and at the same time keep the debate going. If NIH had not acted as rapidly as it did, one can be almost certain that the research would be going on at a much faster pace than it is today, and that the public would know even less than it does now. If NIH had decided to declare a long moratorium on all recombinant DNA research, it would have gained neither the support of the public nor of most segments of the scientific community. NIH could not have made such a moratorium stick.

The Local Debates

Nonetheless, up to and including the February 1976 NIH meeting, there was very little public participation in the recombinant DNA debate. That was soon to change. As the headline of a February 11, 1977, article by Nicholas Wade in *Science* put it, "Gene-Splicing: At Grass-Roots Level a Hundred Flowers Bloom." New York State held public hearings in the fall of 1976 on recombinant DNA research, and a bill has been introduced into the New York legislature to regulate the research. The same thing has happened in California and is being considered in New Jersey. In Madison, Wisconsin, Bloomington, Indiana, and San Diego, California, discussions and local hearings have taken place. A major debate at the University of Michigan resulted in a 6–1 vote among the Regents allowing the research there to proceed. The most flamboyant squabble took place in Cambridge, Massachusetts, where the city council, led by Mayor Alfred Vellucci, imposed a moratorium on the building of a recombinant DNA laboratory at Harvard. The moratorium was lifted only after a special citizen's commission approved the construction.

Meanwhile, a number of environmental groups also became actively involved with the issue during 1976. The Environmental Defense Fund, the Natural Resources Defense Council, Friends of the Earth, and the Sierra Club have all, in one way or another, taken an exceedingly dim view of the way the public has been involved in the debate and have, with minor variations, called for a brand new public debate. Senator Dale Bumpers (D-Ark.) has introduced a bill in Congress to regulate recombinant DNA research, and Senator Edward Kennedy (D-Mass.) is expected to hold hearings this spring before the Senate Health Subcommittee.

Questions for the Public

In one way or other, the public is now involved. But a few questions remain, and I want to focus on three that seem to me central.

1. Why did it take the public so long to get involved in the issue? It does not seem adequate to say that it was because scientists kept the public out. They did not. Even if it is true that wide press coverage is not public participation, it is surely a necessary condition for that participation. The Asilomar Conference was well-covered in the press, there were frequent follow-up stories throughout 1975 and, as the debate heated up, even more in 1976. If the public had wanted to jump in right after Asilomar, it could have done so. But the public did not leap at the opportunity. The lag-time between a complex scientific issue being raised in public and public interest in that issue is almost always fairly long. Yet I think a more subtle, and supplementary, explanation is necessary in this case. My own theory is that the public did not take a real interest in the issue until some senior and notable scientists entered the fray, and entered it in a very outspoken way on the side of the doubters. Erwin Chargaff, George Wald, Ruth Hubbard, and Liebe Cavalieri entered the debate during 1976. To be sure, [Robert] Sinsheimer had gone on record earlier with his own doubts, and Science for the People risked and got considerable wrath for publicizing their own hesitations. But none of them commanded the kind of public attention that Chargaff, Wald, Hubbard and Cavalieri did—if only because of their scientific eminence and seniority.

Why was that important? It is very hard, if not impossible, for the public to get interested in scientific decisionmaking unless potential social and ethical issues are called to their attention by scientists. It is even more difficult to mount a full public debate unless the public has some scientists to lead them into battle. So it was with the early environmental struggles and the debates over nuclear power plants. As Mayor Vellucci said in New Times (in his characteristically understated way), "If I'm gonna take a stand against this goddamn thing, I need some people on my side. And since they [Hubbard and Wald] said they would come, I was fortified, I was ready for a meeting, and that is the reason why we then flung the challenge at Harvard and MIT to send their scientists over here because I knew I had scientists on my side!" Whether their worries and arguments were right or wrong, the presence of Chargaff and others provided a scientific rallying point for those members of the public who wanted to know if their own hesitations had any scientific basis.

2. Now that the public is involved, what options are open to it and how might they be evaluated? A number of options have been proposed, which can be classified into two groups. The first and most moderate would be to turn the present NIH guidelines into state or federal law, perhaps modified, perhaps not. The advantage of that approach is that it would make up for the most obvious and necessary deficiency in the NIH regulations: they apply only to federal grantees. They do not apply to those doing recombinant DNA research under private grants and, most important, they do not apply to private industry. Moreover, they are only guidelines, lacking the force of statutory law. In my own view, there should be such laws, and preferably federal rather than state laws. More to the point a public debate over whether there should be laws governing recombinant DNA research would have the healthy effect of allowing significant public participation.

The other options would go considerably farther. They would mean scrapping the present guidelines and beginning the whole debate over again. The Environmental Defense Fund and the Natural Resources Defense Council have petitioned the DHEW [Department of Health, Education and Welfare] for hearings to determine if any recombinant DNA research should be allowed to go forward, and under what conditions. Friends of the Earth goes still farther, demanding a moratorium on hazardous experiments pending the outcome of Congressional hearings. Wright has called for a complete moratorium on all recombinant DNA research "until policy options have been carefully considered and chosen through democratic procedures developed for the purpose." Sinsheimer, Wald, and Chargaff would allow the research to be done in only one national laboratory, and then under very strict controls. Clifford Grobstein of the University of San Diego has offered a middle-ground proposal that would establish a joint commission, appointed by the President and Congress. Its task would be to take up the ethical, social, and legal issues earlier skirted as well as to analyze all aspects of the problem. A full assessment by the Commission would be due not later than two years from its initiation and not more than three years from the date of the NIH guidelines.

I do not think a full and total moratorium is possible, even if it might be desirable. Such a flat ban on research would require a consensus which simply does not exist. It would also beg the question of whether the research is potentially hazardous. Nor, for that matter, does there exist any consensus on the need for caution and slow movement which would have to be the underlying value premise of such a moratorium. I also have doubts about concentrating all research in a single facility. Is it the best way to get the best science? At least we want that. Will such a necessarily well-guarded and quarantined facility be a good starting point for full public disclosure of all the facts as the research moves forward? Full disclosure is more likely where many facilities are doing the research and where many scientists, involved in the research or not, are able to watch the process. It stretches my imagination to suppose that the sharpest critics of the research would be offered positions in a one-and-only national recombinant DNA facility.

Will it be possible to develop the democratic mechanisms for full public debate now being called for? I believe so, first by means of the debate necessary to turn the NIH guidelines into federal law, making them applicable to all who do the research; and, second, by the establishment of a federal commission to fully examine the issues, or, alternatively, to make use of the present National Commission for the Protection of Human Subjects. In the meantime, speaking as at least one member of the public, I am well prepared to live with the present NIH guidelines.

3. What ethical and social criteria should the public use in judging and deciding upon the future of recombinant DNA research? This is to me the most fundamental question. Though I think that the course chosen by the original group which signed the letter to the National Academy and the Institute of Medicine, and by the NIH in developing its guidelines, was, given the novelty of the issue and the need for quick action, eminently defensible, the critics are still correct in a very general way. There has yet to be a good national public discussion. Worse still, the discussion seems not to be getting any better. It is, in fact, becoming boringly repetitive in substance and tediously hysterical in tone. It is simply not enough to affirm the high principle that the public should have a role in the decision-making, that there should be public forums and public debates. What ought to be the content of that discussion? By what criteria ought the public to judge the competing scientific and ethical cases that have been made by now? For the public should not only be heard. The public ought also to think. But what is it supposed to think about?

First, the public needs to think most carefully about the whole idea of scientific progress. As a general policy, does it favor boldness and risk-taking, or does it favor caution and low risk? Which is the wisest future direction of our public policy? And what counts as wise? Second, does the public, after considerable thought, think that our society has a moral obligation to pursue lines of research which may benefit present and future generations? I say "moral obligation" because it is sometimes implied by advocates of recombinant DNA research that science would be guilty of a sin of omission if it did not continue and promote research so promising in theoretical and practical benefits. I would prefer to say that the research is desirable and valuable, but by no means is it morally obligatory. It is just one choice among many we can make in allocating our scientific resources. But I would like to know what the public—after due consideration—thinks about all that. Third, what does the public think about risks and benefits? How, in some rational way, ought the public to think about that problem?

One obvious implication of this line of thinking is that the public has as much obligation to act responsibly as does the scientific community. The calls for socially responsible scientists could well be matched with some concern about a socially responsible public. The future of the recombinant DNA debate will depend on the quality of the dialogue between the scientific community and the public. Neither side can conduct the debate on its own. The public must be kept informed in the future, must have a central role in present policy formulation, and must develop standards by which to judge the issues. Scientists must bring their knowledge and, just as important, their lack of knowledge out into the open, not just once but again and again.

The public and the scientific community have now begun to talk. This marriage can be saved.

Source: Daniel Callahan, "Recombinant DNA: Science and the Public," *Hastings Center Report*, April 1977, 20–23.

DOCUMENT 6: National Institutes of Health Guidelines for rDNA Research (1976)

The NIH guidelines are lengthy and complex. They detail who can do such research, the conditions under which it can be carried out, who must review it, and how it is to be conducted. These guidelines form the core of the guidelines for rDNA research at all facilities supported by federal funds. These selections focus on security requirements for the two highest levels of containment, BL3 and BL4.

* * *

Appendix G-II-C-4. Laboratory Facilities (BL3)

Appendix G-II-C-4-a. The laboratory is separated from areas which are open to unrestricted flow within the building. Passage through two sets of doors is the basic requirement for entry into the laboratory from access corridors or other laboratories or activities may be provided by a double-doored clothes change room (showers may be included), airlock, or other access facility which requires passage through two sets of doors before entering the laboratory.

Appendix G-II-C-4-b. The interior surfaces of walls, floors, and ceilings are water resistant so that they can be easily cleaned. Penetrations in these surfaces are sealed or capable of being sealed to facilitate decontaminating the area.

Appendix G-II-C-4-c. Bench tops are impervious to water and resistant to acids, alkalis, organic solvents, and moderate heat.

Appendix G-II-C-4-e. Each laboratory contains a sink for hand washing. The sink is foot, elbow, or automatically operated and is located near the laboratory exit door.

Appendix G-II-C-4-f. Windows in the laboratory are closed and sealed.

Appendix G-II-C-4-g. Access doors to the laboratory area containment module are self-closing.

Appendix G-II-C-4-h. An autoclave for decontaminating laboratory wastes is available preferably within the laboratory.

Appendix G-II-C-4-i. A ducted exhaust air ventilation system is provided. The system creates directional airflow that draws air into the laboratory through the entry area. The exhaust air is not recirculated to any other area of the building, is discharged to the outside, and is dispersed away from the occupied areas and air intakes.

Appendix G-II-D. Biosafety Level 4 (BL4)

Appendix G-II-D-2. Only persons whose presence in the facility or individual laboratory rooms is required for program or support purposes are authorized to enter. The supervisor has the final responsibility for assessing each circumstance and determining who may enter or work in the laboratory. Access to the facility is limited by means of secure, locked doors; accessibility is managed by the Principal Investigator, Biological Safety Officer, or other person responsible for the physical security of the facility. Before entering, persons are advised of the potential biohazards and instructed as to appropriate safeguards for ensuring their safety. Authorized persons comply with the instructions and all other applicable entry and exit procedures. A logbook signed by all personnel indicates the date and time of each entry and exit. Practical effective protocols for emergency situations are established.

Appendix G-II-D-2-d. Personnel enter and exit the facility only through the clothing change and shower rooms. Personnel shower each time they exit the facility. Personnel use the air locks to enter or exit the laboratory only in an emergency.

Appendix G-II-D-2-e. Street clothing is removed in the outer clothing change room and kept there. Complete laboratory clothing (may be disposable), including undergarments, pants and shirts or jump suits, shoes, and gloves, is provided and used by all personnel entering the facility. Head covers are provided for personnel who do not wash their hair during the exit shower. When exiting the laboratory and before proceeding into the shower area, personnel remove their laboratory clothing and store it in a locker or hamper in the inner change room. Protective clothing shall be decontaminated prior to laundering or disposal.

Appendix G-II-D-2–1. Laboratory animals involved in experiments requiring BL4 level physical containment shall be housed either in cages contained in Class III cabinets or in partial containment caging systems, such as Horsfall units, open cages placed in ventilated enclosures, or solid-wall and -bottom cages placed on holding racks equipped with ultraviolet irradiation lamps and reflectors that are located in a specially designed area in which all personnel are required to wear one-piece positive pressure suits.

Appendix G-II-D-4-b. Walls, floors, and ceilings of the facility are constructed to form a sealed internal shell which facilitates fumigation and is animal and insect proof. The internal surfaces of this shell are resistant to liquids and chemicals, thus facilitating cleaning and decontamination of the area. All penetrations in these structures and surfaces are sealed. Any drains in the floors contain traps filled with a chemical disinfectant of demonstrated efficacy against the target agent, and they are connected directly to the liquid waste decontamination system. Sewer and other ventilation lines contain high efficiency particulate air/HEPA filters.

Source: National Institutes of Health Guidelines for rDNA Research, 1976.

DOCUMENT 7: Nicholas Wade, "Genetics: Conference Sets Strict Controls to Replace Moratorium" (1975)

The first Asilomar conference established a year-long moratorium on rDNA research to give time to study the safety issues. By the next year's conference many scientists were ready to terminate the moratorium and get on with the research. This selection presents the views of James Watson, who pushed for resuming research because he saw the risks involved in rDNA research as less dangerous than the risks to which he and colleagues were exposed to in work he was already doing on tumor viruses.

* * *

The next event was when Watson got up and said he thought the moratorium should end. This was surprising because Watson had been a member of the committee that asked for it. The reason, he explained, was that "when we met I thought we should have 6 months to see if we could hear anything that would frighten us. As someone in charge of a tumor virus laboratory, I feel we are working with something which is instinctively more dangerous than anything I have heard about here. . . . The dangers involved are probably no greater than working in a hospital. You have to live with the fact that someone may sue you for $1 million if you are careless. That sounds very negative and right-wing but I don't see any other way of doing it."

Watson was speaking for one side of an important divide in contemporary molecular biology, the cancer virologists and old-style microbiologists who are used to dealing with highly infectious agents and to whom, for example, such habits as shutting off washroom faucets with their elbows instead of hands were evidently second nature. Several of them spoke with horror of the "sloppiness" and "prostitution of microbiological technique" of the younger molecular biologists who have recently invaded the field but still treat viruses and bacteria as just another bench reagent.

Source: Nicholas Wade, "Genetics: Conference Sets Strict Controls to Replace Moratorium," *Science*, March 14, 1975, 931–34.

DOCUMENT 8: Allen Silverstone, Testimony to Cambridge, Mass., City Council with Respect to the Establishment of Guidelines for rDNA Research in Cambridge, Mass. (1976)

This document presents one perspective on the technology of rDNA research as it was debated during attempts to draft legislation for such research in the city of Cambridge, Massachusetts.

* * *

First of all, our concern is not abstract. The members of our groups are among the first to be exposed, and like workers in vinyl chloride plants, are the guinea pigs for the safety (or lack) of the new technology. The danger of these new agents, as described earlier, is unique. Any hazard associated with these agents can be spread like any bacterial infection . . . it can multiply—so it is not something to be trifled with. Unless we can be assured that the possibility of danger is reduced to insignificance, we would suggest that the NIH withhold funding such research, until the questions of safety and procedure are settled, especially to the satisfaction of honest critics within the scientific community, and the public. We do not propose this lightly. We recognize that many scientists wish to do these experiments, and thus far their restraint has been admirable.

In all facilities, from the P1 to P4 level, there will be a distinct possibility of cross contamination of cultures. Even the most expert microbiologists, using the most sophisticated equipment, will contaminate a culture. And whether the agent is E. coli or mycoplasma, or what have you, the danger is clear. No concept of safety within a facility can be centered on one experiment (or, on one experimenter). The possibility of multiple interaction is likely, and truly dangerous.

Thus I think we must suggest the possibility of cutting the funding of health and safety violators, and rejecting the publication of experiments that do not observe safety practice (although perhaps publication and public censure would be better). We believe, however, it is inappropriate for the peer review process to become tied up in the question of evaluating safety. . . . We would rather see a committee that evaluated separately from scientific merit, the proposed safety features of an experiment.

Source: Allen Silverstone, Representative of MIT Biological Workers Health and Safety Committee, Testimony to Cambridge, Mass., City Council with respect to the establishment of guidelines for rDNA research in Cambridge, Mass., 1976.

DOCUMENT 9: Stanley N. Cohen, "Recombinant DNA: Fact and Fiction" (1977)

One of the debates that accompanied the development of various DNA technologies considered whether potential biohazards would accompany the application of these technologies and whether any benefits from these technologies might outweigh such potential problems. This document argues that the risk of biohazards is small and that the benefits significantly outweigh potential risks.

* * *

Almost 3 years ago, I joined with a group of scientific colleagues in publicly calling attention to possible biohazards of certain kinds of experiments that could be carried out with newly developed techniques for the propagation of genes from diverse sources in bacteria (1).

Guidelines have long been available to protect laboratory workers and the general public against known hazards associated with the handling of certain chemicals, radioisotopes, and pathogenic microorganisms; but because of the newness of recombinant DNA techniques, no guidelines were yet available for this research. My colleagues and I wanted to be sure that these new techniques would not be used, for example, for the construction of streptococci or pneumococci resistant to penicillin, or for the creation of Escherichia coli capable of synthesizing botulinum toxin or diphtheria toxin. We asked that these experiments not be done, and also called for deferral of construction of bacterial recombinants containing tumor virus genes until the implications of such experiments could be given further consideration.

During the past 2 years, much fiction has been written about "recombinant DNA research." What began as an act of responsibility by scientists, including a number of those involved in the development of the new techniques, has become the breeding ground for a horde of publicists—most poorly informed, some well-meaning, some self-serving. In this article I attempt to inject some relevant facts into the extensive public discussion of recombinant DNA.

Some Basic Information

Recombinant DNA research is not a single entity, but rather it is a group of techniques that can be used for a wide variety of experiments. Much confusion has resulted from a lack of understanding of this point by many who have written about the subject. Recombinant DNA techniques, like chemicals on a shelf, are neither good nor bad per se. Certain experiments that can be done with these techniques are likely to be hazardous (just as certain experiments done with combinations of chemicals taken from the shelf will be hazardous), and there is universal agreement that such recombinant DNA experiments should not be done. Other experiments in which the very same techniques are used—such as taking apart a DNA molecule and putting segments of it back together again—are without conceivable hazard, and anyone who has looked into the matter has concluded that these experiments can be done without concern.

Then, there is the area "in between." For many experiments, there is no evidence of biohazard, but there is also no certainty that there is not a hazard. For these experiments, guidelines have been developed in an attempt to match a level of containment with a degree of hypothetical risk. Perhaps the single point that has been most misunderstood in the controversy about recombinant DNA research is that discussion of "risk" in the middle category of experiments relates entirely to hypothetical and speculative possibilities, not expected consequences or even phenomena that seem likely to occur on the basis of what is known. Unfortunately, much of the speculation has been interpreted as fact.

There is nothing novel about the principle of matching a level of containment with the level of anticipated hazard; the containment procedures used for pathogenic bacteria, toxic substances, and radioisotopes attempt to do this. However, the containment measures used in these areas address themselves only to known hazards and do not attempt to protect against the unknown. If the same principle of protecting only against known or expected hazards were followed in recombinant DNA research, there would be no containment whatsoever except for a very few experiments. In this instance, we are asking not only that there be no evidence of hazard, but that there be positive evidence that there is no hazard. In developing guidelines for recombinant DNA research, we have attempted to take precautionary steps to protect ourselves against hazards that are not known to exist—and this unprecedented act of caution is so novel that it has been widely misinterpreted as implying the imminence or at least the likelihood of danger.

Much has been made of the fact that, even if a particular recombinant DNA molecule shows no evidence of being hazardous at the present time, we are unable to say for certain that it will not devastate our planet some years hence. Of course this view is correct; similarly, we are unable to say for certain that the vaccines we are administering to millions of children do not contain agents that will produce contagious cancer some years hence, we are unable to say for certain that a virulent virus will not be brought to the United States next winter by a traveler from abroad, causing a nationwide fatal epidemic of a hitherto unknown disease—and we are unable to say for certain that novel hybrid plants being bred around the world will not suddenly become weeds that will overcome our major food crops and cause worldwide famine.

The statement that potential hazards could result from certain experiments involving recombinant DNA techniques is akin to the statement that a vaccine injected today into millions of people could lead to infectious cancer in 30 years, a pandemic caused by a traveler-borne virus could devastate the United States, or a new plant species could uncontrollably destroy the world's food supply. We have no reason to expect that any of these things will happen, but we are unable to say for certain that they will not happen. Similarly we are unable to guarantee that any of man's efforts to influence the earth's weather, explore space, modify crops, or cure disease will not carry with them the seeds for the ultimate destruction of civilization. Can we in fact point to one major area of human activity where one can say for certain that there is zero risk? Potentially, we could respond to such risks by taking measures such as prohibiting foreign travel to reduce the hazard of deadly virus importation and stopping experimentation with hybrid plants. It is possible to develop plausible "scare scenarios" involving virtually any activity or process, and these would have as much (or as little) basis in fact as most of the scenarios involving recombinant DNA. But we must distinguish fear of the unknown from fear that has some basis in fact: this appears to be the crux of the controversy surrounding recombinant DNA.

Unfortunately, the public has been led to believe that the biohazards described in various scenarios are likely or probable outcomes of recombinant DNA research. "If the scientists themselves are concerned enough to raise the issue," goes the fiction, "the problem is probably much worse than anyone will admit." However, the simple fact is that there is no evidence that a bacterium carrying any recombinant DNA molecule poses a hazard beyond the hazard that can be anticipated from the known properties of the components of the recombinant. And experiments involving genes that produce toxic substances or pose other known hazards are prohibited.

How About the Benefits?

For all but a very few experiments, the risks of recombinant DNA research are speculative. Are the benefits equally speculative or is there some factual basis for expecting that benefits will occur from this technique? I believe that the anticipation of benefits has a substantial basis in fact, and that the benefits fall into two principal categories: (i) advancement of fundamental scientific and medical knowledge, and (ii) possible practical applications.

In the short space of $3\frac{1}{2}$ years, the use of the recombinant DNA technology has already been of major importance in the advancement of fundamental knowledge. We need to understand the structure and function of genes, and this methodology provides a way to isolate large quantities of specific segments of DNA in pure form. For example, recombinant DNA methodology has provided us with much information about the structure of plasmids that cause antibiotic resistance in bacteria, and has given us insights into how these elements propagate themselves, how they evolve, and how their genes are regulated. In the past, our inability to isolate specific genetic regions of the chromosomes of higher organisms has limited our understanding of the genes of complex cells. Now use of recombinant DNA techniques has provided knowledge about how genes are organized into chromosomes and how gene expression is controlled. With such knowledge we can begin to learn how defects in the structure of such genes alter their function. On a more practical level, recombinant DNA techniques potentially permit the construction of bacterial strains that can produce biologically important substances such as antibodies and hormones. Although the full expression of higher organism DNA that is necessary to accomplish such production has not been achieved in bacteria, the steps that need to be taken to reach this goal are defined, and we can reasonably expect that the introduction of appropriate "start" and "stop" control signals into recombinant DNA molecules will enable the expression of animal cell genes. On an even shorter time scale, we can expect recombinant DNA techniques to revolutionize the production of antibiotics, vitamins, and medically and industrially useful chemicals by eliminating the need to grow and process the often exotic bacterial and fungal strains currently used as sources for such agents. We can anticipate the construction of modified antimicrobial agents that are not destroyed by the antibodies inactivating enzymes responsible for drug resistance in bacteria.

In the area of vaccine production, we can anticipate the construction of specific bacterial strains able to produce desired antigenic products, eliminating the present need for immunization with killed or attenuated specimens of disease-causing viruses.

One practical application of recombinant DNA technology in the area of vaccine production is already close to being realized. An E. coli plasmid coding for an entire toxin fatal to livestock has been taken apart, and the toxin gene has been separated from the remainder of the plasmid. The next step is to cut away a small segment of the toxin-producing gene so that the substance produced by the resulting gene in E. coli will not have toxic properties, but will be immunologically active in stimulating antibody production.

Other benefits from recombinant DNA research in the areas of food and energy production are more speculative. However, even in these areas there is a scientific basis for expecting that the benefits will some day be realized. The limited availability of fertilizers and the potential hazards associated with excessive use of nitrogen fertilizers now limits the yields of grain and other crops, but agricultural experts suggest that transplantation of the nitrogenase system from the chromosomes of certain bacteria into plants or into other bacteria that live symbiotically with food crop plants may eliminate the need for fertilizers. For many years scientists have modified the heredity of plants by comparatively primitive techniques. Now there is a means of doing this with greater precision than has been possible previously. Certain algae are known to produce hydrogen from water using sunlight as energy. This process potentially can yield a virtually limitless source of pollution-free energy if technical and biochemical problems indigenous to the known hydrogen-producing organisms can be solved. Recombinant DNA techniques offer a possible means of solution to these problems.

It is ironic that some of the most vocal opposition to recombinant DNA research has come from those most concerned about the environment. The ability to manipulate microbial genes offers the promise of more effective utilization of renewable resources for mankind's food and energy needs; the status quo offers the prospect of progressive and continuing devastation of the environment. Yet some environmentalists have been misled into taking what I believe to be an antienvironmental position on the issue of recombinant DNA.

An inevitable consequence of these containment procedures [specified by the NIH] is that they have made it difficult for the public to appreciate that most of the hazards under discussion are conjectural. Because in the past, governmental agencies have often been slow to respond to clear and definite dangers in other areas of technology, it has been inconceivable to scientists working in other fields and for the public at large that an extensive and costly federal machinery would have been established to provide protection in this area of research unless severe hazards were known to exist. The fact that recombinant DNA research has prompted international meetings, extensive coverage in the news media, and governmental intervention at the federal level has been perceived by the public as prima facie evidence that this research must be more dangerous than all the rest. The scientific community's response has been to establish increasingly elaborate procedures to police itself—but these very acts of scientific caution and responsibility have only served to perpetuate and strengthen the general belief that the hazards under discussion must be clear cut and imminent in order for such steps to be necessary.

It is worth pointing out that despite predictions of imminent disaster from recombinant DNA experiments, the fact remains that during the past 3½ years, many billions of bacteria containing a wide variety of recombinant DNA molecules have been grown and propagated in the United States and abroad, incorporating DNA from viruses, protozoa, insects, sea urchins, frogs, yeast, mammals, and unrelated bacterial species into E. coli, without hazardous consequences so far as I am aware. And the majority of these experiments were carried out prior to the strict containment procedures specified in the current federal guidelines.

Despite the experience thus far, it will always be valid to argue that recombinant DNA molecules that seem safe today may prove hazardous tomorrow. One can no more prove the safety of a particular genetic combination under all imaginable circumstances than one can prove that currently administered vaccines do not contain an undetected self-propagating agent capable of producing cancer in the future, or that a hybrid plant created today will not lead to disastrous consequences some years hence. No matter what evidence is collected to document the safety of new therapeutic agents, a vaccine, a process, or a particular kind of recombinant DNA molecule, one can always conjure up the possibility of future hazards that cannot be disproved. When one deals with conjecture, the number of possible hazards is unlimited; the experiments that can be done to establish the absence of hazard are finite in number.

Those who argue that we should not use recombinant DNA techniques until or unless we are absolutely certain that there is zero risk fail to recognize that no one will ever be able to guarantee total freedom from risk in any significant human activity. All that we can reasonably expect is a mechanism for dealing reasonably with hazards that are known to exist or which appear likely on the basis of information that is known. Beyond this, we can and should exercise caution in any activity that carries us into previously uncharted territory, whether it is recombinant DNA research, creation of a new drug or vaccine, or bringing a spaceship back to Earth from the moon. Today, as in the past, there are those who would like to think that there is freedom from risk in the status quo. However, humanity continues to be buffeted by ancient and new diseases, and by malnutrition and pollution; recombinant DNA techniques offer a reasonable expectation for a partial solution to some of these problems. Thus, we must ask whether we can afford to allow preoccupation with and conjecture about hazards that are not known to exist, to limit our ability to deal with hazards that do exist. Is there in fact greater risk in proceeding judiciously, or in not proceeding at all? We must ask whether there is any rational basis for predicting the dire consequences of recombinant DNA research portrayed in the scenarios proposed by some. We must the examine the "benefit" side of the picture and weigh the already realized benefits and the reasonable expectation of additional benefits, against the vague fear of the unknown that has in my opinion been the focal point of this controversy.

Note

1. P. Berg, D. Baltimore, H. W. Boyer, S. N. Cohen, R. W. Davis, D. S. Hogness, D. Nathans, R. Roblin, J. D. Watson, S. Weissman, N. D. Zinder, *Proc. Natl. Acad. Sci. U.S.A.* 71, 2593 (1974).

Source: Stanley N. Cohen, "Recombinant DNA: Fact and Fiction," *Science* 195 (February 18, 1977): 654–657.

DOCUMENT 10: Jeremy Rifkin, "DNA: Have the Corporations Already Grabbed Control of New Life Forms?" (1977)

Jeremy Rifkin is a long-standing critic of many technological innovations in genetics and biology. In this selection he focuses on rDNA research conducted in the private sector, where it will be outside the scope of federal regulations because the funding is private. Although safety is an issue for Rifkin, more important for him is control of the technology.

* * *

At this moment, microbiologists are at work in more than 180 separate laboratories across the country, busily spending more than 20 million dollars in government grants in pursuit of the creation of new forms of life. They are experimenting with so-called recombinant DNA. By now most newspaper readers have heard of the controversy surrounding DNA research at universities. But, sheltered from the glare of publicity that bathes every new debate at Harvard or Stanford, something more ominous is happening. Today seven major drug companies are engaged in, or about to begin, recombinant DNA research. The companies will soon apply for patents on the new forms of life they are developing. In time this research will translate into an unparalleled commercial bonanza for the pharmaceutical, chemical and agricultural companies as they introduce literally dozens of new life-form products into the market place.

How does one even begin to look at a technology that could eventually lead to the creation of new plants, animals and even the alteration of the human species?

And then there is a more immediate question before us as we enter the Organic Age: should our present corporate system be used as the developing and marketing process when life itself is the product?

Source: Jeremy Rifkin, "DNA: Have the Corporations Already Grabbed Control of New Life Forms?" *Mother Jones*, February/March 1977, 22–23, 26–39.

DOCUMENT 11: James S. Hall, "James Watson and the Search for Biology's 'Holy Grail' " (1990)

James Watson was the co-discoverer, along with Francis Crick, of the structure of DNA. Here we find him profiled, together with discussions of new developments and research in genetics. Also emphasized are a variety of ethical and social issues that have accompanied developments in genetics: privacy, prenatal diagnosis, and genetic screening as a condition for insurance or employment, for example. Finally, the latest phase of genetics research, the Human Genome Project, is highlighted.

* * *

Arguably the best-known biologist in America, winner of the Nobel Prize in 1962, famous for his bestselling science memoir, *The Double Helix*—infamous for its breezy, abrasive style—Watson has played a major role in shaping the direction of biological research since the 1950s. He had absolutely nothing to prove when, in 1988, National Institutes of Health director James Wyngaarden asked him to coordinate NIH's effort in its massive and controversial project. So why did he agree to do it?

Watson considers the question only briefly. He is no longer the skinny, leering lad of the archival photographs, with the exaggerated grin and flip of hair straight out of the Social Realists, cousin to one of Edward Hopper's loners or Thomas Hart Benton's earnest Midwestern string beans. The hair is gray and thinner after all these years, but still wild, tufted like pulled cotton; the voice soft, yet hectic. "No one else wanted to do it," he replies, arching his eyebrows. "And someone had to."

That's not the real reason, of course. Since that first summer at Cold Spring Harbor, Watson has instinctively drifted toward the key questions of biology, and for the past 35 years those questions have danced around the maypole of the double helix. The genome project promises a kind of culmination: complete explication of the molecule. So he commutes to NIH headquarters in Bethesda, Maryland, two days a week and wows Congressional committees with his outspoken, oddly patriotic view of the initiative, and has assumed highly visible leadership of a project that will cost about $3 billion, take 15 years and, it is argued, revolutionize biomedicine in the 21st century.

The genome project promises great strides in the diagnosis and treatment of disease, but that sunny picture trails clouds of heavy social weather: the possibility of eugenics, of abuses of genetic privacy, of human tragedy inherent in diagnosing fatal diseases without being able to treat them. Without question, the project will take abstract medical, legal and ethical dilemmas and dump them upon nearly every hearth and home. Ironically, the only scientist with enough power and visibility to influence the ethical climate in which this new genetic future is forged may be James Watson. So when he says, "I may have to visibly take the heat," he might be alluding not only to the current controversies, but ultimately to how history views his overall stewardship of the project.

"You really want to get the problem solved"

"His dominant moral value," says one associate, "is good science." As a good scientist, Watson believes, "You really are curious about the laws of nature, and you want people to understand them. If you can help in that, that's great. But you really want to get the problem solved and not your own reputation furthered." He pauses to give a short laugh. "I enjoy myself most," he says, "in the company of people whom I find brighter than myself, rather than the other way around." Even more than place, home for Watson has always been the company of such people.

He was born on April 6, 1928, into what he describes as a "poor family" on the South Side of Chicago. His Episcopalian father was a bill collector ("It was an awful job," Watson says; "he would have been better as a schoolteacher, but he never ended up that way"). His Scotch-Irish mother, a Catholic, worked as a secretary at the University of Chicago and served as a Democratic precinct boss. Watson's childhood was shaped by the Great Depression, not just its poverty but its embrace of public service, its New Deal populism, and its model of forceful and charismatic leadership in the person of Franklin Roosevelt, whose name often pops up in Watson's conversation. But one of his earliest conflicts, no doubt, was reconciling that populist sentiment with the estranging brilliance of his own intellect. As a child, he was small, unathletic, bookish and, one gathers, as prickly and undiplomatic as he is known to be today. "I wasn't a popular kid," he recalls, "and I suspect it was because I would generally say something which I thought was true. In those days, I used to think manners were terrible, you know. The truth was important and manners often hid the truth." He expectorates one of his laughs, as if what he has said is so obvious there was hardly any reason to say it at all.

He sought the company of adults and the comfort of adult activities, especially reading. His father, a voracious reader, also introduced his son to bird-watching. With his mother, Watson argued the relative importance of nature versus nurture (young Watson, a self-described "leftist," recalls that he argued the side of environment, while his mother championed heredity).

As a youngster, Watson competed on the Quiz Kids radio show; he lasted three sessions before tripping up, he claims, on questions about Shakespeare and the Old Testament ("If I'd known the religious questions," he has said, "my father would have been angry at me"). He attended the University of Chicago high school, graduated at age 15 and earned a degree in zoology at the University of Chicago in 1947, before pursuing graduate studies at Indiana University.

Bird-watching acquainted Watson with the beauty of nature; Erwin Schrodinger's book *What Is Life?* introduced him to life's central mystery. Schrodinger described the gene as the crucial entity in the study of biology; unfortunately, in the 1940s no one knew exactly what a gene was, what it looked like or how it worked. As he has done throughout his career, Watson smelled its importance and attached himself to a fraternity of brilliant men—it was always men—hot on the trail of the central question.

The story of the discovery of the structure of DNA is told in unforgettable fashion in Watson's racy and candid memoir, *The Double Helix*. Armed with his PhD, at age 21 he headed off to Europe to find the gene. He worked first in Copenhagen on a Merck fellowship, then grew bored with biochemistry and decided the answer was best pursued at the Cavendish Laboratory in Cambridge, England. His training in genetics complemented the strength of Cambridge, where the British enjoyed a reputation for using x-ray pictures to determine the three-dimensional structure of important molecules, and so he was uniquely positioned to solve the problem. "I was as trained to find the structure of DNA," Watson has since observed, "as Prince Charles is trained to be king."

In Cambridge, Watson teamed up with the 35-year-old Francis H. C. Crick, a scintillating intellect with hardly a credit to his thread bare academic resume. "Jim and I hit it off immediately," Crick wrote in his recent autobiography, *What Mad Pursuit*, "partly because our interests were astonishingly similar and partly, I suspect, because a certain youthful arrogance, a ruthlessness, and an impatience with sloppy thinking came naturally to both of us." Crick noted that "Watson was regarded, in most circles, as too bright to be really sound." These two bad boys of biology—arrogant, opportunistic, behaving "insufferably" in Crick's estimation—puzzled their way to success without performing a single experiment. With the American chemist Linus Pauline as a rival, Watson and Crick used x-ray pictures of DNA, produced by Rosalind Franklin and Maurice Wilkins of King's College in London, to divine the shape of this most central molecule. By the spring of 1953, they had determined that it was a double helix; by 1962, Watson and Crick—with Wilkins— had their Nobel Prize.

The self-confidence to think big

It is tempting to view everything that followed as anti-climax, yet Watson reinvented his role in science in the '60s. He stopped being an experimentalist and started being an impresario, a "mover of people." He grabbed the compass and charted the course of research. He attracted grant money. He recruited young talent, mindful of the soil in which he himself was nourished. "I think it's important that you establish conditions where people will really become important early in life," he says, "because it gives them the self-confidence to think big."

That process began in earnest at Harvard University, where Watson took an assistant professorship in 1955, and continued at Cold Spring Harbor, where he assumed the directorship in 1968. That same year, at age 39, he married Elizabeth Lewis, his 19-year-old laboratory assistant. Watson and his wife moved into the 19th-century Airslie House, a hilltop mansion once owned by Louis Comfort Tiffany, where they have raised their two sons, Rufus and Duncan.

When he sat down to write his account of the discovery of the structure of DNA, Watson began with an enigmatic anecdote about a fellow scientist who apparently snubbed him one day simply by asking, "How's Honest Jim?" It adverts, Watson says, to the belief that Watson and Crick pirated Rosalind Franklin's x-ray data en route to their discovery. "You know, 'Honest Jim' was the name you'd give to a used-car salesman," Watson admits now. Yet he adopted it as the title of the manuscript he began to circulate privately; published in 1968 as *The Double Helix*, its disparaging and sometimes damaging tone caused a firestorm of controversy even as it climbed the best-seller charts.

Watson has been "something of a wild man"

The journal *Science* recently observed, "To many in the biological community, [Watson] has long been something of a wild man, and his colleagues tend to hold their collective breath whenever he veers from the script"; one former Cold Spring Harbor board member remarks that Watson has perfected the use of "petulance as a management tool." But colleagues argue that there is always strategy behind his antagonistic style. Watson himself says, "I would not dare take this on unless I thought I had the support of the major people doing it. I don't make unilateral decisions."

It is precisely Watson's candor and integrity, and his willingness to take the heat, that have earned that support. "If it were not Jim—if it were headed by an ordinary person—then I believe the project would not have acquired the degree of acceptance that it has in the scientific community," says David Botstein, a vice president of Genentech and probably the project's most articulate and longstanding critic. "I'm convinced he's the right person in the sense that he is the person who scientifically carries the most credibility for this project," agrees Sydney Brenner of the Medical Research Council in England. "Jim is actually the heroic figure of science, in terms of the whole development of this area."

To contemplate the enormousness of the task, consider a freckle on the back of your hand. The average freckle contains several thousand cells. Each one of those cells contains 23 pairs of chromosomes. The amount of DNA in each bundle of chromosomes, if unfurled, would form an invisible thread reaching about six feet in length. Interspersed randomly along that six feet of information are somewhere between 50,000 and 100,000 genes. What biologists now propose to do is to locate each one of those genes, pinpoint its specific location on a specific chromosome (a process called mapping) and ultimately decode the biochemical information down to the so-called "letters" of inheritance, the four basic constituents (referred to by their first letters—A, C, G and T) of all genes. In the double helix these letters are linked in pairs, whose sequence on a gene makes it different from any other. These sequences involve 3 billion pairs in all; to date, biologists have deciphered about 35 million.

Yet the genome project commences at a time when there is not enough money to fund much deserving basic research, and Watson seems to view the initiative as an opportunity for biologists to reposition the field. "I think the genome is a way of actually focusing medical research," he says. "I think American biomedical research is in a crisis generated by its own success. There are too many good things to do, and it's been enormously successful from the viewpoint of basic science. But except for the help that it's given to AIDS—finding the virus, getting the test for the blood—it hasn't had any major obvious impacts on too many people's lives. And it isn't curing AIDS. I think that many major diseases will be understood when we can get their genetic basis. And I think we may have to produce another Salk vaccine or something like that, so people can see that the implications are worth it."

Watson has two priorities: to bring the project home under cost and to identify key genes related to human disease. An estimated 4,000 human diseases have a genetic component. Some diseases, like sickle-cell anemia, stem from a single misplaced letter in a single gene; in other illnesses, like hypertension, diabetes, some cancers and heart disease, there is an ensemble of genes playing a dissonant chord together.

Since 1911, when color blindness was linked to the male X chromosome, biologists have plodded through the genome in search of genes linked to human disease, but the work has exploded in the past decade. In August 1989, for example, teams from the University of Michigan and Toronto's Hospital for Sick Children announced the isolation of a gene whose malfunction causes cystic fibrosis, which afflicts 1 in 2,000 newborn children. The discovery exemplifies the power of modern genetics, but also the limitations. Finding the gene does not guarantee finding a cure (the genetic glitch of sickle-cell anemia has been known for 15 years, but no cure is in sight). During the past five years the Cystic Fibrosis Foundation has spent $120 million trying to locate the gene. It is being argued that spending about $200 million a year over a 15-year period for the whole genome will be more cost-effective, and more egalitarian, in locating all human genes.

Watson believes that maladies like schizophrenia, alcoholism and Alzheimer's disease might yield the first fruits of the genome project. "We have to get some real results in the next five years," he says. "I mean, better maps to human disease genes. I think you should have goals like that. Find the gene for something which you might not have found if you didn't have the human genome mapped. If you want to understand Alzheimer's disease, then I'd say you better sequence chromosome 21 as fast as possible. And it's unethical and irresponsible not to do it as fast as possible."

Of all the hats Watson will wear as genome project leader, none would appear more ill-fitting at first, or more important, than that of moral ombudsman. Watson is not visibly comfortable with "the ethics thing," as he once termed it; and when it is suggested that, by virtue of his fame and influence, his opinions on ethical issues will likely carry disproportionate weight with Congress and the public, he recoils from the notion like a child ordered to eat peas. "I can't do that," he demurs, "because I'm not, you know, a real player."

His visibility and credibility, however, make him the major player. Protestations not withstanding, Watson has already taken a surprisingly aggressive fiscal stance to insure that ethical issues get a full public airing. He has declared that 3 percent of the NIH project's budget—over 15 years this portion could approach a substantial $90 million—must be devoted to study and research on the ethical implications of mapping the human genome. That outlay may represent the largest targeted funding of biomedical ethics in the modern era.

"The fact that Watson opens his mouth and utters the word 'ethics' changes the whole valence," says Nancy S. Wexler of Columbia University, who heads NIH's working group on ethics for the genome project. "Other people start to think, 'Gee, this must be important. This is something we should pay attention to.' " Biologist Maynard Olson of Washington University, a member of NIH's Genome Advisory Board, calls Watson's commitment of money for studying ethics "a courageous and precedent-setting position to take." And it is not universally applauded among his colleagues. One of Watson's closest associates observes, "There is a vigorous debate within the genome scientific constituency about the wisdom of supporting ethical, legal and social analysis, and if you took a vote, my guess is that it would be 60–40 or more against Watson. So it is all the more remarkable that he would stick his neck so far out."

Watson has stacked his ethics committee with, if anything, thinkers who have been historically attuned to misuses of scientific knowledge. Its members include Jonathan Beckwith of the Harvard Medical School, a thoughtful critic of genetic technologies; Patricia King, a bioethicist at Georgetown University; Robert F. Murray Jr. of Howard University, an expert on the early abuses of genetic screening for sickle-cell anemia and the potential of such programs to feed racial prejudice; and Nancy Wexler, a neuropsychologist who faces the consequences of genetic screening because Huntington's disease, which is invariably fatal, runs in her family. The group's mission, Wexler says, is to "anticipate now the abuses that can occur and then prevent them before they occur."

Genetic screening is nothing new, as scientists are quick to point out, but the genome project will generate hundreds of disease-related tests, and the sheer volume means that prenatal testing and genetic diagnosis will probably touch every life. Scientists promise a new age of medicine, when doctors will use the genetic profile of newborns to detect genetic susceptibilities to disease and will prescribe preventive steps necessary to forestall illness later in life. But such advance knowledge also invites misuse.

Will employers demand human genetic screening before hiring? Will insurance companies demand genetic profiles of its clients and centralize the information? Will insurers decline to pay medical expenses if a fetus with a lifelong genetic disability is brought to term? Will law enforcement agencies seek access to genetic information? (The FBI already maintains a DNA databank on criminals.) How will doctors counsel a family in which medical science can diagnose a fatal illness, like Huntington's disease, but to whom it can offer no cure? And with a complete catalog of genes at our disposal, will we be able to resist the temptation to tinker with genes, not only to cure disease in individuals (somatic-cell gene therapy) but also to rewire family genomes by retooling germ-line, or reproductive, cells. The ground work for this type of genetic reengineering is already being laid. "We must not shy away from the germ line," declared *Science* editor Daniel Koshland at Human Genome I.

What Watson tells his colleagues about all this may be more important than what he tells an interviewer. Here is what he said in San Diego: "It would be naive to say any of these answers are going to be simple. About all we can do is stimulate the discussion, and essentially lead the discussion instead of having it forced on us by people who say, 'You don't know what you're doing.' We have to be aware of the really terrible past of eugenics, where incomplete knowledge was used in a very cavalier and rather awful way, both here in the United States and in Germany. We have to reassure people that their own DNA is private and that no one else can get at it. We're going to have to pass laws to reassure them. [But] we don't want people rushing and passing laws without a lot of serious discussion first."

Can Watson foresee the misuse of genetic material in our society, perhaps 50 or 100 years down the road? "Sure," he replies. "That's why I'm talking about laws. But I think you see a misuse of everything. You see a misuse of pesticides. You see a misuse of high school athletics. You see a misuse of aspirin, of automobiles. So I don't think it should make us upset to say that there will be a misuse."

And when he is asked about what area in the future may be the ripest for misuse of genome information, Watson pauses a long time, hands clasped over his head, and uncharacteristically sidesteps the question. "I don't have a good feeling for where it could be misused the most," he says. And later: "I think people can't really think about things before they are real."

Here, the scientist vies with the New Deal populist. "We're a democracy and the people finally make their decisions," says the populist. "If some process is regarded as repulsive to a majority of the people, we better know." Pause. "Now," adds the scientist, "one hates to have something discovered that's considered repulsive, unless it's been talked about enough to know whether it really is repulsive."

Watson has a better record at looking forward than he allows. In 1971, in an essay for the *Atlantic* called "Moving Toward the Clonal Man," he warned of the coming dilemmas of reproductive technologies and argued against what he termed the "laissez-faire nonsense" that science inevitably makes the world better. Test-tube babies, in vitro fertilization and surrogate parenting sounded futuristic back then, as some of the genome's potential dilemmas sound in 1990. But as "Honest Jim" wrote in 1971, ". . . if we do not think about it now, the possibility of our having a free choice will one day suddenly be gone."

Source: James S. Hall, "James Watson and the Search for Biology's 'Holy Grail,'" *Smithsonian*, February 1990, 41–49.

Part II

Animal Applications

This section will focus on the application of the new information and capabilities in genetics on animals. There are three broad areas of focus: (1) the development of what are called transgenic animals, animals that have genetic information from a different species incorporated into their own genome; (2) the use of such animals for both research and food products; (3) the issuing of patents on such animals. The general concerns common to these topics are the question of harm to the animals, the appropriateness of the use of such animals for human benefit, the safety of such animals either as food products or as the source of medical products consumed by humans, and the status and integrity of their own genome.

DOCUMENT 12: Rudolf Jaenisch, "Transgenic Animals" (1988)

Transgenic animals are those in which a foreign gene has been inserted. This permits the gene to be both expressed in this organism and passed on to its descendants. These organisms can then be used to study the function of the particular gene as well as to develop animal models to study various diseases. The following document is an excerpt from an article that discusses the possible uses for this type of technology.

* * *

The introduction of genes into the germ line of mammals is one of the major recent technological advances in biology. Transgenic animals have been instrumental in providing new insights into mechanisms of developments and developmental gene regulation, into the action of oncogenes, and into the intricate cell interactions within the immune system. Furthermore, the transgenic technology offers exciting possibilities for generating precise animal models for human genetic diseases and for producing large quantities of economically important proteins by means of genetically engineered farm animals.

For many experimental purposes, it would be highly desirable to be able to modulate expression of a transgene with some external stimulus. Promoters of genes subject to modulation by hormonal or other environmental stimuli have been shown in several instances to properly control expression of transgenes. The metallothionein promoter, for example, has been used to direct expression of many different reporter genes, and in some cases expression could be stimulated by feeding the animals with heavy metals. Hormone-inducible promoters that function in transgenic mice include the mouse mammary tumor virus (MMTV) long terminal repeat (LTR) (35, 36), the transferrin gene (37), the H-2 Eα gene (38), and two liver-specific genes. These results are promising and indicate that transgene expression can be modulated in vivo by external signals. However, at present, the stimuli used to activate inducible genes show toxic side effects that limit their experimental utility.

The ability to introduce and express genes in the animal has opened the door to efforts designed to correct genetic defects. So far "gene therapy," that is, the repair of a mutated gene, has not been accomplished in the animal. Genes introduced into animals invariably integrate at a site distant from the resident defective gene. Therefore, mutant and introduced normal genes will segregate independently in the next generation. Nevertheless, successful corrections of hereditary disorders at the phenotypic level have been accomplished and include hormone deficiencies (40), β-thalassemia (42), and a myelination defect (42). These types of experiments will help to elucidate the molecular deficiency causing the respective hereditary disorder.

Genes have also been microinjected into rabbit, sheep, and pig embryos (43). The success rate of generating transgenic domestic farm animals is, however, much lower than that obtained with mice, in large part because of technical difficulties in visualizing the pronucleus in the embryos of these species. A human growth hormone gene successfully introduced into the pig in spite of these difficulties was nevertheless unable to increase growth, perhaps because the human hormone was not biologically active in pigs (43). Although the importance of genetic engineering for improving livestock has been questioned (44), it is likely that transgenic farm animals may become a source of economically valuable proteins. For example, medically relevant proteins, whose expression has been targeted to the mammary epithelial cells, may be harvested from the milk of transgenic cows as has been shown to be possible for oncogene transgenic mice (45).

It is likely that rapid advances will occur in the following areas. (i) It will be important to isolate and characterize chromosomal regulatory elements controlling developmental gene activation over large distances (31). Inclusion of such elements in gene constructs should guarantee predictable and efficient expression independent of the chromosomal integration site. This will be particularly important for genetic engineering of large farm animals where cost constraints limit the number of transgenic lines that can be generated and evaluated. (ii) The various possibilities of marking early embryonic cells or ablating specific lineages give experimental access to stages of mammalian development as yet not amenable to easy experimental manipulation. This undoubtably will accelerate our understanding of the complex cell interactions in mammalian development. (iii) The prospect for generating recessive or dominant mutations in preselected genes not only will permit the derivation of precise animal models for human hereditary diseases but also will mark the beginning of a systematic genetic dissection of developmental processes that will radically change the future of experimental mammalian genetics.

REFERENCES AND NOTES

31. F. Grosveld, G. van Assendetft, D. R. Kollias. *Cell* 51, 957.

40. R. E. Hammer, R. D. Palmiter, R. L. Brinster, *Nature* 311, 65 (1984); A. J. Mason et al., *Science* 234, 1372 (1986).

41. F. Costantini, K. Chada, J. Magram, *Science* 233, 1192 (1986).

42. C. Readhead et al., *Cell* 48, 703 (1987); B. Popko et al., ibid., p. 713.

43. R. E. Hammer et al., *Nature* 315, 680 (1985); R. Hammer et al., *J. Anim. Sci.* 63, 269 (1986); G. Brem et al., *Zuchthygiene* (Berlin) 20, 251 (1985).

44. R. Land and I. Wilmut, *Theriogenology* 27, 169 (1987).

45. K. Gordon et al., *Biotechnology* 5, 1183 (1987); J. P. Simons, M. McClenaghan, and J. Clark, *Nature* 328, 530 (1987).

Source: Rudolf Jaenisch, "Transgenic Animals," *Science* 240 (June 10, 1988): 1468–1474.

DOCUMENT 13: John Travis, "Scoring a Technical Knockout in Mice" (1992)

When a particular gene is removed from an organism, in the genetic vernacular it is referred to as being "knocked out." This allows researchers to custom design an organism for research. This is important because the absence of the gene in the organism will help researchers understand the gene's function much better. The following document explores this concept.

* * *

Making use of a natural cell process known as homologous recombination, scientists can now almost routinely create strains of mice and other animals in which a selected gene is disrupted—"knocked-out"—in the lab vernacular.

The fruits of this conceptual breakthrough have, in the last year or so, become increasingly obvious. Just a few years ago, data from the knockout mice Capecchi and others were producing entered the scientific literature in a trickle. But now the fields of immunology, cancer genetics, and developmental biology are blessed with a steady stream of results from these designer mutant mice. Indeed, the very success of the work has created a problem as the researchers grapple with how to make these popular animals available to others.

Round One: The Immune System

The field that has so far benefited the most from knockouts is immunology, where the technique is helping to clarify the functions of the myriad genes needed for the immune system's operations (*Science*, 24 April, p. 483). The reason is simple. When some genes are knocked out, the defects that result are so severe that the embryos die very early in development, before researchers can glean much useful information from them.

Round Two: Developmental Biology

As the immune system begins to yield some of its secrets to knockout mice, developmental biologists are also turning to these mutant animals to help solve another fundamental problem in biology: how genes control early development. "For mammalian developmental genetics, this has really opened up the field. It's going to be used all over the place," says Alexandra Joyner of Mount Sinai Hospital in Toronto, whose own group just used a knockout to show that a gene called N-mwc helps control the shaping of lungs in the embryo.

By far the biggest knockout effort in developmental biology focuses on the homeobox genes, so called because they contain an evolutionarily conserved sequence known as the homeobox. These genes, which were identified a few years ago as key players in the embryonic development of fruit flies, are also active in higher animals, including mice and other mammals. But before knockouts became available, researchers had to rely on indirect clues, such as where the gene was expressed during embryogenesis, as they sought to tease out the role of the homeobox genes in mouse development.

Round Three: Genetic Disorders

So far, knockout technology's major contribution has been in basic research, helping researchers decipher the functions of individual genes, but it is beginning to open up a whole new avenue of more applied research: the creation of animal models of human genetic diseases. Cystic fibrosis, sickle-cell anemia, beta thalassemia, and Gaucher's disease are just a few of the genetic disorders that researchers are trying to recreate in mice with knockout technology. Even diseases that appear to involve contributions by multiple genes, like atherosclerosis and Alzheimer's, may reveal their secrets by combining two or more knockouts in the same animal, suggests Smithies.

But even if researchers are ready to face the technical, monetary, and temporal hurdles in pursuit of knockouts, there's a final trap awaiting them. It may take a year or two to create a particular knockout strain, but it can take even longer to understand the pathology of the mice. "The hard part now is to effectively analyze the mutant phenotypes," says Joyner. And the ultimate frustration, of course, is knocking out a gene and seeing no change at all.

Despite the problems, however, researchers are confident that knockouts will become easier. Experienced labs can now make a new knockout in less than a year. And some are even poised to take the next step: They are trying to do "cleaner" gene exchanges, sometimes called the "hit-and-run" method, in which the neo gene is not present in the final gene in the modified stem cells. This would open the way to recreating specific mutations in mice, not just knocking out a gene by disrupting its sequence. Meanwhile, most knockout researchers are ecstatic about the control they already have over the mouse genome. "It's opened up a completely new set of capabilities for understanding mammalian development and gene function," says Smithies.

Source: John Travis, "Scoring a Technical Knockout in Mice," *Science* 256 (June 5, 1992): 1392–1394.

DOCUMENT 14: Novo Nordisk Research Corporation, Statement on the Use of Transgenic Animals in Research (1994)

This statement of an international research company with facilities in the United States and Denmark specifies the benefits of using transgenic animals and also shows how their use conforms to Danish legislation.

* * *

These animals greatly facilitate the search for important new drugs and reduce the number of laboratory animals used.

A transgenic animal that makes a human protein (e.g., human insulin) will recognize this substance as its own and will therefore not produce an immune response against it. As a consequence, the identity and purity of the product can be tested more efficiently in such animals, thereby saving the use of many laboratory animals otherwise needed to obtain a statistically significant result.

Complicated human proteins can often not be produced efficiently and in their correct form by microorganisms or mammalian cell cultures. Some human proteins, however, can be secreted in large amounts and in their correct form in the milk or blood of mammals such as cows, goats, sheep, rabbits, or mice. In certain cases, transgenic animals represent the only means to produce important new drugs.

In Denmark, research involving transgenic animals needs approval by the Ministry of Justice, the Ministry of Labour, and the Ministry of the Environment and Energy.

The Danish Ethical Council on Animals has accepted research on and the use of transgenic animals if the purpose is essential and the animals do not suffer. At a consensus conference in Copenhagen in 1992, the use of transgenic animals was considered permissible if the purpose in each case was considered to be reasonable and if the potential suffering of the animals had been considered in relation to the purpose of the research or use.

Source: Novo Nordisk Research Corporation, Statement on the Use of Transgenic Animals in Research, 1994. http://www.nnbt.com.

DOCUMENT 15: Peter R. Wills, "Transgenics a Threat to Nature" (1995)

This selection presents a critique of the use of transgenic animals and plants particularly with respect to their capacity to undermine current ecological stability. The author argues that the development of new life forms may upset any stability inherent within evolution.

* * *

If there is in and among the DNA of different organisms some distributed encoding of ecological stability and the possibility of non-catastrophic evolutionary change, then biotechnology should be viewed as an enterprise which may scramble the natural algorithm and lead to an unmanageable collapse of biological systems as we know them. No wonder the Parties to the 1992 Convention on Biological Diversity want to discuss a biosafety protocol governing genetically modified organisms. Miller sees no special ecological danger in biotechnology. Rather, like the harbingers of nuclear power generation who promised us electricity so cheap that it wouldn't have to be metered, he promised the developing world "environmentally friendly biotech innovations that can clean up toxic wastes, purify water and displace agricultural chemicals." Unfortunately, the long term side effects of fraying the phylogenetic tapestry cannot be foreseen and will not be considered in any of the risk analyses which Miller advocates as a basis for international regulation. The "Biodiversity Treaty" is one legitimate tool for use in helping to bring human dalliance with genetically modified organisms under some meaningful control. Contrary to what Miller suggests, it can be used to do this without prejudicing the development of instruments to guarantee the safety of activities involving other organisms.

Source: Peter R. Wills, "Transgenics a Threat to Nature," Submitted as a Letter to the Editor, *Nature*, February 16, 1995.

DOCUMENT 16: Gary L. Francione, "Animal Rights and Animal Welfare: Five Frequently Asked Questions" (1996)

Some critics suggest that the way to protect animals is to shift the argument from animal welfare, which most individuals support, to animal rights, which few support. The argument of this selection is that such a shift would make harming animals much more difficult.

* * *

Our current legal system currently reflects a welfarist approach, and it clearly does not work. The law recognizes that animals have interests in being treated "humanely" or in being kept free from "unnecessary" suffering. These laws require that we "balance" human interests against these animal interests; despite such laws, we still have pigeon shoots, facial branding, castration without anesthesia, circuses, rodeos, etc. These uses of animals are completely "unnecessary" and "inhumane" as these terms are used in ordinary language, but they are all protected under the law.

The reason for this failure to protect animals is found in the legal status of animals as the property of human beings. Animals may have interests, but these interests may be traded away or sacrificed even when the primary reason for sacrificing the interest is a completely trivial "benefit" in the form of human amusement and entertainment. As animals are regarded as property, it is almost always in some human's interest to exploit those animals.

For example, the 1985 amendments to the federal Animal Welfare Act, which created animal care committees to ensure the "humane" treatment of animals used in biomedical experiments, recognize very explicitly that animals have interests, but then permit their use for virtually any purpose as long as experimenters consider it "necessary." The continued use of animals in painful and bizarre experiments indicates that virtually any animal interest provided under the Act can be outweighed by any human interest, including the mere curiosity of the vivisector. Laws such as the Animal Welfare Act and the federal Humane Slaughter Act are generally ineffective, except for convincing those people sitting on the fence that it is all right to exploit animals because we do it so thoughtfully.

Source: Gary L. Francione, "Animal Rights and Animal Welfare: Five Frequently Asked Questions" (Rutgers Animal Rights Law Center, Newark, NJ 1996).

DOCUMENT 17: Philip Leder and Timothy A. Stewart, Transgenic Non-Human Mammals (1988)

Here we have a selection from the first patent granted for a transgenic non-human mammal, Patent No. 4,736,866. The patent presents in both drawings of the gene sequence in question and in descriptions of the process and its uses how the mouse was developed and to what uses it may be put. The language is highly technical, but this selection gives a flavor of the patent.

* * *

Abstract: A transgenic non-human eukaryotic animal whose germ cells and somatic cells contain an activated oncogene sequence introduced into the animal, or an ancestor of the animal, at an embryonic stage. . . .

An activated oncogene sequence, as the term is used herein, means an oncogene, which, when incorporated into the genome of the animal, increases the probability of the development of neoplasms (particularly malignant tumors) in the animal. . . .

Introduction of the oncogene sequence at the fertilized oocyte stage ensures that the oncogene sequence will be present in all of the germ cells and somatic cells of the transgenic animal. The presence of the oncogene sequence in the germ cells of the transgenic "founder" animal in turn means that all of the founder animal's descendants will carry the activated oncogene sequence in all of their germ cells and somatic cells. . . .

The animals of the invention can be used to test a material suspected of being a carcinogen, by exposing the animal to the material and determining neoplastic growth as an indicator of carcinogenicity. This test can be extremely sensitive because of the propensity of the transgenic animals to develop tumors. This sensitivity will permit suspect materials to be tested in much smaller amounts than the amounts used in current animal carcinogenicity studies, and thus will minimize one source of criticism of current methods, that their validity is questionable because the amounts of the tested material used is greatly in excess of amounts to which humans are likely to be exposed. . . .

The animals can also be used as tester animals for materials, e.g., antioxidants such as betacarotine or Vitamin E, thought to confer protection against the development of neoplasms. . . .

Source: Philip Leder and Timothy A. Stewart, Transgenic Non-Human Mammals, U.S. Patent 4,736,866, April 12, 1988.

DOCUMENT 18: Robert Wachbroit, "Eight Worries about Patenting Animals" (1988)

This document presents an overview of eight issues the author considers problematic with respect to the practice of patenting animals.

* * *

In April the United States Patent and Trademark Office issued patent No. 4,736,866 to Harvard University for a genetically engineered cancer-prone mouse; the same patent would cover any species of transgenic mammal containing the patented recombinant cancer sequence. This, the world's first patent for an animal, has sparked intense controversy and aroused vociferous objections.

We may divide the objections to animal patenting into intrinsic versus extrinsic objections. An intrinsic objection finds patenting animals itself to be morally wrong; it refers to values or ideals that are associated with the very idea of an animal patent. Intrinsic objections turn more on what a patent claim symbolizes than on what power or control it confers, since most people who object to animals as intellectual property do not object to animals as personal property. Extrinsic objections concern the likely consequences of animal patents; these arguments are therefore based on empirical or contingent assumptions about what these likely consequences would be. Extrinsic objections often express broader worries about biotechnology in general.

Here are the various arguments against patenting animals that have surfaced in public debate, first the intrinsic, then the extrinsic objections.

1. Patenting animals is incompatible with the sanctity of life. The paradigm of patentable subject matter is the mechanical invention. If life is seen along this dimension—and so, as Jeremy Rifkin, a noted critic of biotechnology, has suggested, an animal is seen to be little more than a complicated toaster—how can we attach any special value to life? Central to viewing life as having a sanctity is viewing the distinction between living and non-living as having a moral component. If there is no moral difference between living and non-living, how could the mere fact of life itself have moral value? According to the leading Supreme Court decision in this area, a living organism is no different from a "composition of matter," as far as patentability is concerned. Viewing organisms as "mere composites of matter" seems to deny that there is a moral difference between living and non-living.

It might be instructive to look at the controversies that surround

Darwin's views on man. Darwin denied certain differences between people and other living things, and some people took this to be a denial of a moral difference between people and animals. But, plainly, it is not. No serious philosopher would argue that the fact of evolution undermines the moral difference between people and animals. Similarly, a materialist or a reductionist view of life needn't be incompatible with there being a moral difference between living and non-living.

Thus, insofar as this worry is a worry about viewing life as having a material basis, it is not a serious moral objection. However, there is more to this objection. This more has to do with the idea that patenting licenses our viewing these organisms as property that is made but not born. Some philosophers would argue that living things are distinguished by their having a certain kind of history rather than by their having a certain kind of composition. In viewing transgenic animals as patentable subject matter, we are identifying them by their chemical makeup, rather than by their origin. The result is to force apart two concepts—that of being alive and that of being born (and not made)—that have been thought to be inseparable. The worry is that our understanding of the sanctity of life cannot be sustained if we divorce these two concepts.

If the details of this objection can be properly filled out, it constitutes a serious objection to patenting animals, unless we can find some satisfactory way of reconstructing our understanding of the sanctity of life.

2. Patenting animals encourages a disrespect for nature. This objection can be understood in an intrinsic or in an extrinsic version.

The intrinsic version goes like this: Our relationship to nature is that of a caretaker or steward, not that of an owner; an owner can do a lot more to his property than a caretaker can to his trust. As caretakers, it is argued, our treatment of nature is subject to constraints against violating the integrity of the "natural order." A patent claim, as a claim of ownership, is thus a falsification of our proper relationship to nature.

The extrinsic version goes like this: It is a small step from thinking that existing species should be protected or that the integrity of a particular environment should be maintained to thinking that the distinction between species should be maintained. Patenting, however, will encourage the invention of creatures that undermine the integrity of the natural order.

The main problem for either version is that of extending environmental values and concerns to domestic and laboratory animals. It seems to me that the extension of environmental concerns to non-wildlife can only be based on false analogies. As they stand, these objections do not raise a serious moral problem for animal patents insofar as transgenic animals are understood as domestic or laboratory animals and not as "products of nature."

3. Patenting encourages a dangerous technology. A primary purpose

of granting patents is to encourage technology and invention. Since there are doubts about the safety of biotechnology, we should establish a moratorium on animal patents until the risks of biotechnology are more fully studied and better understood.

Proponents of patenting are quick to reply that granting a patent does not certify safety. A patent does not grant a right to make, use, or market an invention; it confers only a right to exclude others from making, using, or marketing the invention. It is inappropriate to use the patent system for regulatory purposes.

This reply, however, may smack of legalism. While it might be important to separate agencies and offices whose mission is to promote or encourage technology from those whose mission is to regulate it, the effects of their actions are not so easily separated. A failure to regulate can be a form of promotion; a failure to encourage can be a form of regulation. The patent system has the power to regulate technology. Furthermore, there is a precedent for using the patent system as a regulatory tool: you cannot obtain a patent for an invention that is useful only in connection with atomic weapons.

All this suggests that a patent for a transgenic animal should not be granted if the animal poses a comparable kind of danger. Certainly the Harvard mouse does not fall into this category. Thus the objection may be a sound one, but only when applied on a case-by-case basis.

4. Patenting animals will encourage inhumane treatment of animals. Arguments over the ethics of animal experimentation have usually turned on whether the benefits to humans sufficiently outweigh whatever harm experimental animals suffer. If we weigh the good against the bad consequences of using transgenic animals in biomedical research, the good consequences seem to dominate. At least this seems clear in the case of the Harvard mouse.

Transgenic animals introduce a further complication, however. Is humane treatment of a mouse humane treatment of a transgenic mouse? If we assume that what counts as humane treatment of an animal depends in part on the kind of animal in question, then at some point the genetic modification of the animal could be so great that the "parent" stock would no longer serve as a guide to what humane treatment consists in. To that extent, the objection makes the valid point that we should proceed cautiously—but as it stands, it does not support a blanket prohibition on animal patents.

5. The patenting of animals will eventually lead to the patenting of humans. Taken literally, this is an absurd objection. No one thinks that allowing animals as someone's personal property constitutes a slippery slope that ends in allowing people as personal property. Why should we think there is a slippery slope from the patenting of animals to the patenting of people?

Nevertheless, there is a deeper worry at work here. While introducing human DNA into a mouse doesn't turn the mouse into a person, it does raise the question of the moral category in which transgenic animals belong, especially those containing human genetic material. What are we to think when such techniques are used on "higher" animals, e.g., monkeys? How ought we to regard such a "halfway" creature? As an augmented ape or as a diminished human?

Such decisions needn't arise, as long as we restrict genetic manipulation so as not to result in any "halfway" animals. But as biomedical research demands better and better animal models, there will be increasing pressure to produce halfway creatures as models for human disorders. Do the benefits of these accurate models outweigh the moral ambiguities and confusions that the creation of these animals would generate? This could become a serious worry. Thus while this argument does not raise an objection to the transgenic mouse, it does raise an objection to using the Harvard patent to create a transgenic primate.

6. Patenting threatens the norms of scientific research. The scientific community's search for truth involves the prompt dissemination of discoveries through publications and professional meetings. If scientists wait for patent approvals before announcing discoveries and if other researchers must obtain license in order to utilize and build on these discoveries, then research suffers. This objection seems especially relevant when, as in this case, the patented item has more value for research than for commercial use.

The usual approach in this topic is to draw a distinction between basic and applied research. The goal of basic research is knowledge; the goal of applied research is a commercial product. In basic research, it is important that scientists can replicate one another's results and build on them in order to advance our knowledge. In applied research, these concerns do not govern, and so applied research can be subject to proprietary concerns. Patenting basic research items is untenable: it doesn't generate any greater incentive to undertake basic research and could obstruct the flow of knowledge, slow down scientific progress, and undermine the cohesion and collegiality of the scientific community.

One could reply that although patents for basic research items may not increase the incentives to do basic research, they increase the financial support for such research. But does this extra benefit outweigh the costs—in particular, the accompanying loss of public support for basic research? Insofar as we value the traditional norms of basic research, there should be a research exemption for patents.

7. The public should not pay twice for the products of publicly funded research. When the basic research underlying a patent has been publicly funded, granting a patent in effect makes the public pay twice for an invention and allows private concerns to profit unfairly.

Those who defend current patent policy may respond that the public interest isn't served by the mere existence of an invention. The invention must also be suitably manufactured, distributed, and marketed. A useful invention gathering dust on the shelf does no one any good. Thus, even if the public has funded research crucial to an invention, patent incentives may still be necessary to ensure that the invention reaches its market.

Yet insofar as there is a suitable market for a product, it seems that there should already be an incentive for private investment in it. The public should invest in research only when its benefits cannot be captured by the pricing mechanism. Perhaps, then, the public should not be supporting the kind of research that is captured so easily in market prices and, therefore, by private investors. Since the public has supported the relevant research, however, it has rights to the information that results.

8. Patenting animals will have an unacceptable economic impact, leading to the economic downfall of the family farm. Owners of small and medium-size farms will not be able to compete against the large agribusinesses that will use more efficient genetically engineered animals, or they will become caught in a technological treadmill, forced to invest more and more heavily in genetic technology simply to remain competitive.

This argument reflects more a political worry than an ethical one. What is wrong with allowing small farmers to go the way of the village blacksmith? There do not seem to be moral reasons for giving special economic protection to the small farmer, although there may be other reasons having to do with our particular culture and history. An assessment of these, however, would take us too far afield.

The objections thus far raised to the patenting of animals are a mixed bag of intrinsic and extrinsic objections, some easily dismissed, others sound and compelling. All need closer philosophical scrutiny as the wisdom and ethics of patenting higher life forms continue to be the subject of public debate.

Source: Robert Wachbroit, "Eight Worries about Patenting Animals," *Report from the Institute for Philosophy and Public Policy* 8, no. 3 (summer 1988): 6–8.

DOCUMENT 19: Mark Sagoff, "Animals as Inventions: Biotechnology and Intellectual Property Rights" (1996)

This selection suggests a way to resolve problems concerning patenting raised by some religious leaders. Whereas some religious leaders are concerned about the impropriety of owning God's creation, chief executive officers are concerned about a legitimate profit on their investment.

* * *

Since 1988, The U.S. Government has granted nine patents for genetically engineered animals. A coalition of religious leaders called last summer for a moratorium on such patents. They declared that "the gift of life from God, in all its forms and species, should not be regarded solely as if it were a chemical product subject to genetic alteration and patentable for economic benefit."

For its part, the biotech industry argues that without patents, companies would be unwilling to invest in, or unable to attract capital for, research that benefits humankind.

There seems to be room for compromise between the concerns of the industry and those of religious leaders. Activist Jeremy Rifkin emphasized that they had "no problem" with patents nor even with protecting biotech products with the sort of marketing exclusivity conferred on "orphan" drugs.

On the industry side, spokespersons have been eager to assure their clerical critics that what they want is not to upstage the Creator but to enjoy a legal regime that protects and encourages investment.

Novel organism might be covered by a new patent statute which, like the Wild Plant Protection Acts, recognizes that those who produce these organisms may alter or recombine but do not design and therefore can not claim to own them as intellectual property.

Source: Mark Sagoff, "Animals as Inventions: Biotechnology and Intellectual Property Rights," *Report from the Institute for Philosophy and Public Policy* 16 (winter 1996).

DOCUMENT 20: Research and Production Issues: National Pork Producers Council Position Statement (1997)

Many patent supporters claim that patents encourage creativity and protect and enhance financial investment. This statement from the National Pork Producers Council (NPPC) states its position on the patenting of animals, particularly of pigs, in a way that reflects its economic concerns with respect to its product.

* * *

Many are concerned that producers could be forced to pay expensive royalties to large corporations that hold the patents on transgenic animals. However, the denial of patent protection could concentrate new development in the hands of only those companies that are financially capable of risking their research investment. Lack of patent protection can clearly be expected to discourage future investment in gene transfer technology.

The option of deliberately delaying technological progress to avoid anticipated social and economic problems would not exist for very long in any part of the world if others refuse to follow this same strategy. At the same time, we acknowledge a need for a regulatory framework that can oversee the new technologies and offer the public the necessary re-assurance that risks and benefits have been duly considered. However, any regulations should be no less and no more than those governing U.S. competitors.

NPPC recognizes the need for biotechnology companies to have suffi-cient incentives to continue developing state-of-the-art research that will benefit the U.S. livestock industry. Patent protection is obviously neces-sary to encourage this kind of meaningful research.

Thus, we believe it is appropriate to allow royalties to be collected on the sale of germ plasm, semen and embryos of patented animals. First generation animals could also be patented, but we recommend farm ani-mals be exempted beyond the first generation. We need to be certain that the existence of patents on transgenic farm animals does not disrupt existing marketing systems for farm animals.

Source: Research and Production Issues: National Pork Producers Council Posi-tion Statement, August 1997.

DOCUMENT 21: Alan R. Shuldiner, M.D., "Transgenic Animals" (1996)

Various transgenic technologies are applied in research. The follow-ing document explores how transgenics are produced and then how such animals, typically mice, are used in research. The article also notes that such technology helps set the stage for somatic gene therapy in humans. This form of therapy cures a disease either by repairing a damaged gene or by replacing the damaged gene with a correct copy of the gene.

* * *

In this article, I will describe how transgenic animals are produced and used in biomedical research. I will focus on the introduction of exogenous DNA into a fertilized egg for the subsequent expression of a protein product. Modifications of this method are now being used to disrupt, or knock out, the expression of endogenous genes—a topic that will be covered in a subsequent article.

The first step in producing transgenic animals is to construct the DNA to be transferred—the transgene—so that the desired gene product will be expressed in the desired location. This step is accomplished with conventional recombinant DNA techniques. Typical transgenes contain nucleotide sequences that correspond to the gene of interest, with all the components necessary for efficient expression of the gene, including a transcription-initiation site, the 5' untranslated region, a translation-initiation codon, the coding region, a stop codon, the 3' untranslated region, and a polyadenylation site. A critical component of the transgene is the promoter, or regulatory region, that drives transcription. The transgene can be expressed in many tissues of the transgenic animal by using a promoter from a ubiquitously expressed gene, such as that of β-actin or simian virus 40 T antigen. Alternatively, with tissue-specific promoters, the transgene can be expressed in a given location. For example, selective expression in fat cells is possible with the adipocyte P2 promoter, and expression in muscle can be accomplished by using the myosin light-chain promoter; the amylase promoter confines expression to the acinar pancreas, and the insulin promoter restricts expression of the transgene to islet beta cells.

Introduction of the transgene into the mouse genome requires fertilized mouse eggs. Injection of gonadotropins (typically, a mixture of pregnant-mare serum gonadotropin and human chorionic gonadotropin) into a female mouse induces hyperovulation, which, followed by natural mating with a fertile male, provides the source of the eggs. The fertilized eggs are harvested before the first cleavage and placed in a petri dish. The DNA construct (usually about 100 to 200 copies in 2 pl of buffer) is introduced by microinjection through a fine glass needle into the male pronucleus—the nucleus provided by the sperm before fusion with the nucleus of the egg. The diameter of the egg is 70 μ and that of the glass needle is 0.75 μm; the experimenter performs the manipulations with a binocular microscope at a magnification of 200X. The injected eggs are cultured to the two-cell stage and then implanted in the reproductive system of a female mouse. A total of 25 to 30 injected embryos are usually implanted. After 19 to 20 days of gestation, pups are born.

Typically, 15 to 30 percent of the injected embryos will proceed to term, and 10 to 20 percent of these full-term embryos will have integrated the transgene into their germ-line DNA. These transgenic pups (called founders) are identified by testing their genomic DNA (usually obtained from the tails of the pups) for the transgene by Southern blot analysis or the polymerase chain reaction. Typically, 1 to 200 copies of the transgene are incorporated in a head-to-tail orientation into a single random site in the mouse genome. Since injection and integration occur before the first cell division, all cells of the founders, including the germ cells, will be heterozygous for the transgene.

It is critical to determine whether the transgene is being expressed in the appropriate location. This requires a quantitative assay of the tissue (or tissues) of interest for the messenger RNA or protein (or both) that corresponds to the transgene. The level of expression of the transgene, which can vary greatly among transgenic animals, depends on many factors, including the intrinsic efficiency of the transgene promoter, the number of copies of the transgene that were incorporated, and to some extent, the site of integration within the mouse genome.

Once expression of the transgene has been confirmed, heterozygous founders are mated with nontransgenic mice, and the resulting heterozygous siblings are mated with each other to generate mice that are homozygous for the transgene. These animals constitute a transgenic line. Transgenic lines that have incorporated different numbers of copies of the transgene at different integration sites are usually produced and studied to establish that a given phenotype is indeed due to over-expression of the transgene and not to artifacts of the manipulations.

Transgenic animals have been used for simulating diseases and testing new therapies. For example, transgenic animals engineered to over-express a mutant form of the gene for β-amyloid protein precursor (the APP gene) have neuropathological changes very similar to those in persons with Alzheimer's disease. This model supports a primary role of the APP gene in the development of Alzheimer's disease, an idea that heretofore was controversial. The model also provides an opportunity to test methods for the prevention or delay of Alzheimer's disease.

Transgenes that express cytotoxic molecules (so-called toxigenes) have been used to extirpate specific tissues. In one such model, a transgene was created in which the gene for diphtheria toxin was placed under the control of a promoter that drives expression specifically in brown adipose tissue. Expression of diphtheria toxin resulted in the extirpation of the tissue. These animals were extremely obese and had defects in their mogenesis—results that support an important role of brown adipose tissue in the regulation of energy balance. The animals also had insulin resistance and diabetes mellitus, making them an appropriate model for studying the relations among obesity, insulin resistance, and glucose intolerance.

A promising new animal model for determining the extent of DNA damage (mutagenesis) caused by chemical toxins has been developed. The model exploits the transgenic expression of LacZ, a bacterial enzyme whose activity can be measured in vitro with a simple colorimetric assay.

LacZ transgenic mice are exposed to the putative mutagen. Relevant tissues are then harvested, DNA is prepared from them, and the LacZ transgene is cloned into a plasmid vector and expressed in bacteria. The frequency of mutations in the LacZ transgene, which corresponds to the number of bacterial colonies that lack LacZ activity, is correlated with the mutagenic capability of the chemical given in vivo. This model has important applications in toxicology, studies of carcinogenesis, research on the role of DNA damage and repair in various diseases, and studies of the normal process of aging.

These are just a few examples of applications of transgenic models in biomedical research. They illustrate the in vivo extension of recombinant DNA technology for the study of specific molecules in diseases and for the production of animal models that can accelerate the discovery of new treatments. Transgenic technology helps set the stage for somatic gene therapy in humans.

RECOMMENDED READING

Brinster R L, Palmiter R D. Introduction of genes into the germ line of animals. *Harvey Lect* 1984–85, 80: 1–38.

Gossen J A, de Leeuw W J, Vijg J. LacZ transgenic mouse models: their application in genetic toxicology. *Mutat Res* 1994; 307:451–9.

Lowell B B, Susulic V, Hamann A, et al. Development of obesity in transgenic mice after genetic ablation of brown adipose tissue. *Nature* 1993; 366:740–2.

Pattengale P K, Stewart T A, Leder A, et al. Animal models of human disease: pathology and molecular biology of spontaneous neoplasms occurring in transgenic mice carrying and expressing activated cellular oncogenes. *Am J Pathol* 1989; 135:39–61.

Source: Alan R. Shuldiner, M.D., "Transgenic Animals," *New England Journal of Medicine* 334 (March 7, 1996): 653–655.

DOCUMENT 22: Baruch Brody, "On Patenting Transgenic Animals" (1995)

This document reviews several of the arguments against patenting animals, particularly the theological and philosophical arguments. Baruch Brody concludes that these arguments are ultimately not successful because they do not necessarily yield the conclusion drawn from them or because the premises of the arguments are too narrowly stated. Brody also argues that the arguments supporting patenting can provide a strong moral defense of the practice.

* * *

Introduction

This paper critically evaluates the arguments the opponents presented. It concludes that the opponents failed to make their case and that the proponents of the patenting of transgenic animals can offer strong moral defense of their position.

Moral Arguments Resting Upon Metaphysical and/or Theological Assumptions

The most fundamental moral arguments opposing patenting employ metaphysical and theological claims as their point of departure. They raise questions about the relation between the living and the non-living, and about the relation between human beings and the world which they inhabit. Moral conclusions are drawn at the end of these arguments, but they begin with moral philosophical and theological claims.

There is one feature of these arguments that needs to be noted. The opponents themselves recognize that they need to do a lot more work to articulate the inchoate concerns they feel. This is why the recent statement of some religious leaders against animal patenting called for a moratorium on patenting while a process of "thoughtful reflection and judgment on these matters by churches and religious institutions, as well as by other concerned groups in our society" is carried out. All that we can do is examine the articulations of concern that have already been developed. We need to be sensitive, however, to the possibility that further articulations of reasons for opposing patenting may be forthcoming.

Shortly after the *Diamond v. Chakrabarty* decision in which the Supreme Court ruled that a living microorganism was patentable, Leon Kass published an important essay in which he raised a number of fundamental questions about the patenting of living organisms of any size or complexity. Here, we focus on one argument which he stated as follows: "Consider first the implicit teaching of our wise men, that a living organism is no more than a composition of matter, no different from the latest perfume or insecticide. What about other living organisms—goldfish, bald eagles, horses? What about human beings? Just compositions of matter? Here are deep philosophical questions to which the Court has given little thought, but in its eagerness to serve innovation, it has, perhaps unwittingly, become the teacher of philosophical materialism." (Kass, 1985: pp. 149–50)

Reinforcing Kass's point was the fact that the majority in *Diamond v. Chakrabarty* was required to find that the organism was a composition of matter, because the statute authorizing patents refers to "any new and useful process, machine, manufacture, or composition of matter" as a patentable object, and the relevant microorganism only fell under the last phrase. This aspect of the decision was also the basis of criticism of it by a working party of the World Council of Churches (1982):

"The U.S. Supreme Court decision on patenting of life forms rested upon a specific, highly reductive conception of life, which sought to remove any distinction between living and non-living matter that could serve as an obstacle to the patenting of living but unnatural organisms." (p. 47) This is obviously not the place to examine the great philosophical debate between materialists, who see living objects as nothing more than complexly organized matter, and non-materialists, who insist that living objects are more than that. Let us leave that question unresolved. Let us go further and agree with Kass that it would be inappropriate for society to adopt a social policy that committed us as a society to a materialistic conception of life. Still, there is nothing in the decision to patent living things even under current language of the patenting statute that commits us to a materialistic conception of life. Even those who believe that living beings are more than compositions of matter believe that they are at least compositions of matter, and it is only as compositions of matter that we patent them. A second argument, that some of the genetic engineering being patented confuses what must be kept distinct, is found in a passage in the recent statement of religious leaders against animal patenting. They make the following claim:

"The combining of human genetic traits with animals, with the results to be patented and owned, raises unique moral, ethical, and theological questions, such as the sanctity of human worth, which must be examined." A good example of what they have in mind is, of course, the introduction of genes for the human growth hormone into farm animals to produce greater growth.

The sanctity of human worth is, of course, a fundamental moral principle of our society, standing behind our belief that humans cannot be killed or mistreated, are entitled to freedom from enslavement, and so forth. Since we allow animals to be killed for food and to be owned, we do not subscribe to a similar sanctity of animal worth principle. After all, a sanctity of worth principle would seem to imply at least the following two elements: (1) the life of the entity in question is of sufficient value that it can be taken only in the most extreme circumstances (e.g. self-defense); (2) the individual is free to act as it desires, for it should not be treated as a mere means for others to attain their ends. By killing animals for use as food, we show that we do not ascribe such significance to their lives. By allowing them to be owned by those who would raise them for use as food, as a source of various byproducts (e.g. wool), as objects to be entered into competitions, or even as pets, we show that we are willing to treat animals as mere means to human ends. All of this is, of course, perfectly compatible with insisting that unnecessary animal suffering should be eliminated. So our moral life, as currently constituted, rests upon that distinction between humans and animals. It would seem, then, that the religious leaders see such genetic experiments as imperiling the belief in the sanctity of human worth by breaking down the barrier between humans and other animals.

Do they? If it were possible (and it is not possible either now or in the foreseeable future) to alter genetically animals so that they had more of those capacities and features (e.g. the capacity to form moral judgments or the capacity to experience the beautiful and sublime) which we see as distinctive to humans, then we would face difficult moral questions as to how such creatures should be treated and as to whether we can continue to maintain a sharp divide between humans and other animals. But, of course, none of these issues is raised by farm animals who grow more because they have a gene that leads them to produce human growth hormone. Nor would they be raised by any of the genetic alterations of animals that will be produced in the foreseeable future. So in what ways do these experiments raise questions about the sanctity of human worth? And in what way does the patenting of their results imperil that belief?

We turn finally to an interconnected series of arguments about man's control over nature, man's responsibility toward nature, and the need to preserve species and protect their integrity which are probably the leading cause of metaphysical and theological disquiet about the patenting of transgenic animals and about the genetic engineering it will promote. A powerful statement of this set of issues is contained in the following testimony of the Rev. Wesley Granberg-Michaelson, appearing on behalf of the National Council of Churches, on November 5, 1987:

"When the National Council of Churches has issued this statement of concern, it comes from the background of Judeo-Christian thinking about how we relate to the natural environment. In a nutshell that background says that we have a responsibility for preserving the integrity of creation, and for working with it in order to preserve its intrinsic values. The doctrine of trust in legal parlance is synonymous to what we are talking about theologically or religiously when we think about the relationship of the creation to humanity. The Judeo-Christian view says that the creation is, in essence, held in trust; there are limitations on what we can do. We have a responsibility to see that its integrity is preserved. This background has led to legislation such as endangered species laws, animal welfare laws, laws regarding environmental quality." (Committee on the Judiciary, 1988)

There are several points which I want to make about this argument: (1) The presentation of Judeo-Christian tradition is somewhat misleading. As John Passmore showed in his ground-breaking study, *Man's Responsibility for Nature* (1974), the traditional Judeo-Christian image was that of man's dominion over nature. To give but one example, Calvin repeatedly talked about the fact that God created all things for man's sake. It is only in recent years that the theme of man's stewardship has become more predominant.

(2) The traditional idea of a steward or of a trustee is the idea of a person who manages property for the benefit of other persons (present and future) who are its owner. There is nothing in the traditional conception of stewardship or trusteeship which even suggests that the property is to be managed to preserve its integrity for its own sake. Property held in trust can be radically transformed by trustees if it serves the best interest of its human owners, present and future. One religious notion of stewardship is the notion that man must treat the property he owns as a trust for those human beings who will follow in future generations and cannot over-exploit it so as to maximize his current benefit. This notion of stewardship is analogous to the legal notion of trusteeship, but it is not the notion that Granberg-Michaelson is employing. He is using the different notion of the steward who protects the integrity of the property for the property's sake. (3) This radically new notion may be morally desirable, but its claims demand considerable justification, and only those who are prepared to accept its many radical implications are entitled to use it as the basis for arguing against transgenic animals and the patenting of them. This argument is analogous to some of the arguments used by some animal rights advocates in that it is based upon a radical revision in our metaphysical conception of the relation of human beings to non-human nature, and not merely on the adoption of the traditional notion of stewardship.

CONCLUSIONS

Where then do we stand? We have seen that the moral arguments normally raised against the patenting of transgenic animals fail. At most, they suggest the real need to strengthen our regulatory schemes governing research on animals and the release of transgenic animals into the wild, and the need to reconsider in a general fashion the problem of justice between countries. None of them provide us with moral reasons to oppose the patenting of transgenic animals, unless we also are prepared to adopt radically new approaches to fundamental metaphysical issues surrounding the relation of humanity to the environment and to the world of animals.

There are, moreover, strong moral arguments for allowing the patenting of transgenic animals. The most important of these is the consequentialist claim that such a patenting system promotes beneficial consequences by providing an incentive to create useful inventions.

Source: Baruch Brody, "On Patenting Transgenic Animals," *AG Bioethics Forum* 7 (November 1995): 1 and 5–7.

Part III

Agriculture

Food occupies a central place in the lives of all people. For some, it is the daily search for enough food. For others, it is the quest for the right food. Still others search for the exotic or unusual. But all, no matter their taste or need, desire sufficient quantity of food and and a supply that is safe. The application of biotechnology to various food products intends to help increase the food supply, to develop foods that are blight resistant, that are cost efficient to produce, and that are safe. Yet even given these goals, many criticize the developers and producers of genetically engineered food products for making foods that some claim are potentially harmful. This section will present a variety of claims and counter claims about genetically engineered food products and means that are being taken to ensure their safety.

DOCUMENT 23: Keith Schneider, "Wisconsin Temporarily Banning Gene-Engineered Drug for Cows" (1990)

This excerpt reports on one of the first oppositions to the introduction of a genetically engineered product, one designed to increase milk production. Of interest is that the critique is directed both to health concerns and to the economic survival of dairy farmers.

* * *

In a decision that could spell enormous trouble for the fledgling agricultural biotechnology industry, Gov. Tommy G. Thompson of Wisconsin yesterday approved legislation that temporarily bars the sale or use of a genetically engineered drug for use in dairy cows.

This action is the first prohibition of a product of genetic engineering anywhere in the country, and it sets off another temporary ban of the drug in neighboring Minnesota.

The political and symbolic significance to the agricultural biotechnology industry could be profound, said industry analysts. Since the development of genetic engineering in the early 1970's, companies using the technology to invent new microbes, plants and drugs for use in agriculture have been convulsed by waves of dissent from environmental groups, farmers, and other opponents.

At issue is whether the products will do more to harm farmers, consumers and rural communities than to help them. Developers of the drug say bovine somatotropin is a triumph of technology that will cause cows to produce 10 to 25 percent more milk on less feed, thus raising productivity and profits. Opponents of the drug say it will drive farmers out of business by encouraging overproduction, and they say questions about its effect on human health have not been adequately answered.

Source: Keith Schneider, "Wisconsin Temporarily Banning Gene-Engineered Drug for Cows," *New York Times*, April 28, 1990, A1 and A10.

DOCUMENT 24: Peter Wills, "Biologist Rejects Transgenic Foods" (1995)

In this document a theoretical molecular biologist states his objection to bioengineered food products. Note that the objection is based on opposition to corporate intervention into the world of nature.

* * *

The main thrust of most genetic engineering is to create organisms which evolution could never produce. Genes are routinely transferred across boundaries between taxa, where there is not even a plausible pathway for DNA traffic. Synthetic genes are created and inserted to produce organisms where no plausible sequence of selective pressures is possible.

It's likely that the natural boundaries which limit DNA traffic, and selective pressures that limit the inheritance of mutations, are vital to long-term evolutionary stability. The damage long-term from interference with evolutionary processes could be much worse than our disruption of shorter-term ecological processes. They should not be interfered with for commercial gain.

I personally belong to a group for whom genetic engineering violates a coherent set of values. For me, food consisting of or derived from genetically modified organisms is not substantially equivalent to traditional foods and I don't want to eat it. This has nothing to do with any scientific proof of a detectable effect on me. I object to such food in principle because the reorganisation of global resources for humans on the basis of biotechnological trickery is fundamentally and morally wrong. It is against what I perceive "Nature" to be and I don't support it.

I wish to live in a society which recognises and caters for personal beliefs like mine and does not demand that I capitulate to the agenda of transnational corporations who gain commercially from tinkering with evolution.

Source: Peter Wills, "Biologist Rejects Transgenic Foods," Letter commenting on the Australia New Zealand Position on Genetically Engineered Food, 1995. http://www.envirolink.org/arrs/essays/transgenics.

DOCUMENT 25: Molly O'Neill, "Geneticists' Latest Discovery: Public Fear of 'Frankenfood' " (1992)

This document reports on the new name given to genetically engineered foods, concerns about these products, and some responses to those concerns.

* * *

Industry analysts say that new genetically engineered food products will be rolled out in the next 18 months. It could be the biggest boon to corporate profits since frozen foods were introduced in the 1930's and it could also be a marketing nightmare. Food technologists, the biologists and chemists who develop food-growing and processing techniques, are at the forefront of the image crisis that is kicking up around genetically altered foods like a dust storm over an Idaho potato field.

Scientists have argued both for and against gene splicing in food products. And agricultural economists have discussed a wide range of altruistic applications.

Indeed the deconstruction and reconstruction of DNA chains in edible plants and animals seem to tap a wellspring of modern misgivings. "There's a distrust of technology, distrust of corporate profits, distrust of Government regulatory agencies and general fears about the safety of the food supply," said Dr. Susan K. Harlander, a leading researcher.

"If the public understood the technology, they would understand that part of their emotional reaction is irrational," she said.

The fact that recent public outcry over genetic engineering of food has centered on control as well—consumers are asking for detained labels that give them control over what they buy and eat—was not lost on Dr. Pernet, of the Rouquette Corporation. To him the consumer's cry for control, like the food marketer's attempts to control the image of genetically altered food, echoes the underlying misgivings about genetic engineering in the scientific community.

Source: Molly O'Neill, "Geneticists' Latest Discovery: Public Fear of 'Frankenfood'," *New York Times*, June 28, 1992, A1 and A14.

DOCUMENT 26: David Kessler, Michael R. Taylor, James H. Maryanski, Eric L. Flamm, and Linda S. Kahl, "The Safety of Food Developed by Biotechnology" (1992)

Here, members of the FDA review methods of developing new varieties of food plants and the FDA's approach to their regulation, which is based on a scientifically grounded safety assessment. Such procedures also include testing of food products on animals and the examination of the wholesomeness of such foods.

* * *

Developers introduce hundreds of new varieties of food plants into commerce every year. Most have improved agronomic characteristics, such as higher yield. Varieties are also being developed with enhanced quality characteristics, such as improved nutritional or processing attributes. To develop new varieties, breeders use all the techniques at their disposal to generate ever more advantageous combinations of genetic traits.

Breeding techniques include hybridizations between plants of the same species, between plants of different species, and between plants of different genera; chemical and physical mutagenesis; interspecies and intergeneric protoplast fusions; soma-clonal variation resulting from regeneration of plants from tissue culture; and in vitro gene transfer techniques.

Traditional methods of plant breeding of some major crops have yielded dramatic changes in food composition, including increases in major plant constituents. For example, traditional breeding resulted in the transformation of the kiwi fruit from a small berry native to Asia to the recognizable variety in our grocery stores. Hundreds of similar or more subtle improvements in the agronomic, food processing, or other attributes of food crops have been achieved without any significant adverse impact on the safety of foods (2–4).

Recombinant DNA techniques, which are being used to achieve the same types of goals as traditional techniques, offer plant breeders a number of useful properties. First, any single-gene trait (and, potentially, multi-gene traits) whose chromosomal location or molecular identity is known can be transferred to another organism irrespective of mating barriers. Second, this transfer can be accomplished without simultaneously introducing undesirable traits that are chromosomally linked to the desirable trait in the donor organism. Thus, the techniques have great power and precision. Currently, more than 30 different agricultural crops developed with recombinant DNA techniques are being tested in field trials. With these techniques, food crops are being developed to resist pests and disease, to resist adverse weather conditions, to tolerate chemical herbicides, and to have improved characteristics for food processing and nutritional content. The genes conferring these traits usually encode proteins that are responsible for the new trait or that directly or indirectly modify carbohydrates or fats in the plant to bring about the desired characteristics. In addition, genes encoding antisense messenger RNA have been introduced to decrease gene expression and thereby bring about the desired new phenotype.

FDA Approach to Regulation

The United States today has a food supply that is as safe as any in the world. Most foods predate the establishment of national food laws, and the safety of these foods has been accepted on the basis of extensive use and experience over many years or even centuries. Foods derived from new plant varieties are not routinely subjected to scientific tests for safety, although there are exceptions. For example, potatoes are generally tested for the glycoalkaloid solanine. The established practices that plant breeders use in selecting and developing new varieties of plants, such as chemical analysis, taste testing, and visual analyses, rely primarily on observations of quality, wholesomeness, and agronomic characteristics. Historically, these practices have been reliable for ensuring food safety (2–4).

Because of the limited nature of most modifications likely to be introduced, the FDA would waste its resources and would not advance public health if it were routinely to conduct formal premarket reviews of all new plant varieties. We will require such reviews before marketing, however, when the nature of the intended change in the food raises a safety question that the FDA must resolve to protect public health. The FDA also has a responsibility to provide scientific guidance on how the safety of foods from new plant varieties should be evaluated, regardless of whether formal FDA review and approval are required.

Safety Assessment: Scientific Basis

Our safety assessment approach like that of others (2–4) addresses important food safety issues that pertain to the host plant, donor organisms, and new substances that have been introduced into the food. The host plant is a benchmark for considering modifications that may affect the safety of food derived from new varieties. Potential new substances considered in this safety assessment are proteins, carbohydrates, and fats and oils because these are the substances that will be introduced or modified in the first plant varieties developed by recombinant DNA techniques.

Developers should consider changes in the concentrations or bioavailability of important nutrients for which a food is widely consumed. For example, if a new tomato variety contained no vitamin C, consumers would need to be informed of that fact through appropriate labeling (for example, a change in the common name).

Most plants produce a number of toxicants and antinutritional factors, such as protease inhibitors, hemolytic agents, and neurotoxins, presumably as a means of resisting natural predators. The concentrations of toxicants in most species of domesticated food plants (for example, corn and wheat) are so low as to present no health concern. In others (for example, potato and rapeseed), breeders routinely screen new varieties to ensure that toxicant concentrations are within an acceptable range. In some cases (for example, cassava and kidney bean), proper preparation, such as soaking and cooking, is required to produce food that is safe to eat (2–4).

Additionally, plants, like other organisms, have metabolic pathways that no longer function because of mutations that occurred during evolution. Products or intermediates of some of these pathways may include toxicants. In rare cases, such silent pathways may be activated by the introduction or rearrangement of regulatory elements, or by the inactivation of repressor genes by point mutations, insertional mutations, or chromosomal rearrangements. Similarly, toxicants ordinarily produced at low concentrations in a plant may be produced at higher levels in a new variety as a result of such occurrences.

However, the likelihood of such events occurring in food plants with a long history of safe use is low. The potential of plant breeding to activate or upregulate pathways synthesizing toxicants has been effectively managed by sound agricultural practices, as evidenced by the fact that varieties with unacceptably high levels of toxicants have rarely been marketed (2–4).

Therefore, the toxicants that are of concern in any particular species are those that have been found at unsafe concentrations in some lines or varieties of that species or related species. In many cases, characteristic properties (such as a bitter taste associated with alkaloids) are known to accompany elevated concentrations of specific natural toxicants, and the absence of bitter taste may provide an assurance that these toxicants have not been elevated to unsafe levels; in other cases, analytical or toxicological tests may be necessary.

Toxicants known to exist in the donor related species, or progenitor lines, may be transferred to the new plant variety, for example, during hybridization of a cultivated variety with a wild, poisonous relative. The possibility that donor-derived toxicants could occur in food derived from a genetically modified plant should be considered. One of the questions raised most frequently about the use of recombinant DNA techniques to develop improved food crops concerns the safety for consumption of substances (now primarily proteins, carbohydrates, and fats and oils) that will be introduced into foods such as fruits, vegetables, grains, and their byproducts. Here we discuss the scientific issues pertaining to these substances in food derived from the new plant variety.

Developments Affecting Substances in Food from New Plant Varieties

Proteins. Proteins (including antisense modifications that modulate expression of native proteins) make up the largest group of substances being introduced into food through recombinant DNA techniques. The scientific issues pertaining to proteins that are derived from other food sources, or that are substantially similar to proteins that are derived from food sources, are known toxicity, allergenicity, and dietary exposure. Thousands of proteins have been safely consumed in the human diet. Many foods produce an allergenic response in some individuals. Foods that commonly cause allergenic responses include milk, eggs, fish, crustacean, mollusks, tree nuts, wheat, and legumes. Although only a small fraction of the thousands of proteins in the diet have been found to be allergenic, all known food allergens are proteins. The transfer of proteins from one food source to another might therefore confer on food from the host plant the allergenic properties of food from the donor plant. For example, the introduction of a peanut allergen into corn might make that variety of corn newly allergenic to people ordinarily allergic to peanuts.

In some of these foods, the protein responsible for the allergenicity is known (for example, gluten protein in wheat). In such cases, the precision of methods such as recombinant DNA techniques allows the developer to determine whether the allergenic determinant has been transferred from the donor to the new variety. In many foods, however, the protein responsible for the allergenicity is not known. In these cases, well-designed in vitro tests, such as serological tests, may provide evidence that the suspected allergen was not transferred or is not allergenic in the new variety.

A separate issue is whether any new protein in food has the potential to be allergenic to a segment of the population. At this time, we are unaware of any practical method to predict or assess the potential for new proteins to induce allergenicity.

Uncertainty may exist about the safety for consumption of a protein that has not been a constituent of food previously (or has no counterpart in food that would serve as a basis for comparison of safety). The degree of testing these new proteins should be commensurate with any safety concern raised by the objective characteristics of the protein.

Generally, the function of proteins that have been introduced into food by recombinant DNA techniques is well known, and these proteins are not known to exert toxic effects in vertebrates. If such well-characterized proteins do not exhibit unusual functions, safety testing will generally not be necessary.

However, certain groups of proteins are known to be toxic to vertebrates. These include bacterial and animal toxins, hemagglutinins, enzyme inhibitors, vitamin-binding proteins (avidin), vitamin-destroying proteins, enzymes that release toxic compounds, and selenium-containing proteins (12). For such substances, testing may be the only means available to ensure safety.

Carbohydrates. Developments that affect carbohydrates will often be modifications of food starches, presumably affecting the content of amylose and amylopectin, as well as the branching of amylopectin. Such modified starches are likely to be functionally and physiologically equivalent to starches commonly found in food and thus would not suggest any specific safety concerns. However, if a vegetable or fruit is modified to produce high concentrations of an indigestible carbohydrate that normally occurs at low concentrations or to convert a normally digestible carbohydrate to an indigestible form, nutritional questions may arise.

Fats and oils. Some alterations in the composition or structure of fats and oils, such as an alteration in the ratio of saturated to unsaturated fatty acids, may have significant nutritional consequences or result in marked changes in digestibility. Such changes may warrant labeling that describes the new composition of the substance. Additionally, safety questions may arise as a result of the presence of fatty acids with chain lengths greater than C22, fatty acids with cyclic substituents, fatty acids with functional groups not normally presenting dietary fats and oils, and fatty acids of known toxicity, such as erucic acid.

Nonclinical Safety Testing

Animal feeding trials of foods derived from new plant varieties are not conducted routinely. However, in some cases testing may be needed to ensure safety. For example, substances with unusual functions or that will be new macroconstituents of the diet may raise sufficient concern to warrant testing. Tests could include metabolic, toxico-logical, or digestibility studies, depending on the circumstances.

Developers may also need to conduct tests on the "wholesomeness" of foods derived from new plant varieties as a means of ensuring that the food does not contain high levels of unexpected, acutely toxic substances. Such tests may provide additional assurance to consumers that food developed by new technology is as safe as food derived from varieties already in their grocery stores. However, animal tests on whole foods, which are complex mixtures, present problems that are not associated with traditional animal toxicology tests designed to assess the safety of single chemicals. Potential toxicants are likely to occur at very low concentrations in the whole food, and the tests may therefore be inadequately sensitive to detect toxicants. Efforts to increase the amount of whole food ingested by the test animals in order to increase the sensitivity and attempt to establish a traditional margin of safety (for example, a 100-fold safety factor) may not always be possible. When tests are contemplated, careful attention should be paid to test protocol, taking into account issues such as nutritional balance and sensitivity.

FDA's science-based approach for ensuring the safety of foods from new plant varieties focuses safety evaluations on the objective characteristics of the food: The safety of any newly introduced substances and any unintended increased concentrations of toxicants beyond the range known to be safe in food or alterations of important nutrients that may occur as a result of genetic modification. Substances that have a safe history of use in food and substances that are substantially similar to such substances generally would not require extensive premarket safety testing. Substances that raise safety concerns would be subjected to closer inquiry. This approach is both scientifically and legally sound and should be adequate to fully protect public health while not inhibiting innovation.

REFERENCES AND NOTES

2. International Food Biotechnology Council, *Regul. Toxicol. Pharmacol.* 12 (no. 3) (1990), part 2.

3. Nordic Council of Ministers and The Nordic Council, *Food and New Biotechnology* (Scantryk, Copenhagen, 1991).

4. *Approaches to Assessing the Safety of Crops Developed Through Wide-Cross Hybridization Techniques*, proceedings of a Food Directorate Symposium held 22 November 1989, Ottawa, Canada (Food Directorate, Health Protection Branch, Health and Welfare Canada, Ottawa, 1990).

12. W. G. Jaffe, in *Toxicants Occurring Naturally in Foods* (National Academy of Sciences, Washington, DC, 1973), pp. 106, 128.

Source: David Kessler, Michael R. Taylor, James H. Maryanski, Eric L. Flamm, and Linda S. Kahl, "The Safety of Food Developed by Biotechnology," *Science* 256 (June 26, 1992): 1747–1752.

DOCUMENT 27: U.S. Food and Drug Administration, "Biotechnology of Food" (1994)

This document presents a brief overview of elements in the food safety review considered appropriate by the FDA.

* * *

The 1992 policy is based on existing food law, and requires that genetically engineered foods meet the same rigorous safety standards as is required of all other foods. The law places a responsibility on producers and sellers to offer only safe products to consumers, and provides FDA with the legal tools for enforcement.

Food law requires pre-market approval for food additives, whether or not they are the products of biotechnology. Therefore, FDA's biotechnology policy treats substances intentionally added to food through genetic engineering as food additives if they are significantly different in structure, function, or amount than substances currently found in food. Many of the food crops currently being developed using biotechnology do not contain substances that are significantly different from those already in the diet and thus do not require pre-market approval.

The 1992 policy statement contains detailed scientific guidance for developers of new plant varieties so that they understand the issues FDA considers important. For example, safety assessments of foods derived from new plant varieties include evaluation of

—the purpose or the intended technical effect of the genetic modification;

—the source, function and stable incorporation of introduced genetic material;

—analytical studies to determine whether the genetic modification had any effects on the composition of the food (such as the levels of important nutrients and naturally occuring toxicants);

—the safety of new or modified substances (for example, proteins, carbohydrates, fats, or oils) in the food.

Source: U.S. Food and Drug Administration, "Biotechnology of Food," *FDA Backgrounder*, May 18, 1994.

DOCUMENT 28: U.S. Food and Drug Administration Office of Premarket Approval, "Guidance on Consultation Procedures: Foods Derived from New Plant Varieties" (1997)

This document itemizes the information the FDA requests when a manufacturer submits a food for a safety evaluation.

* * *

The safety and nutritional assessment summary should contain sufficient information for agency scientists to understand the approach the firm has followed in identifying and addressing relevant issues. Such information would ordinarily include:

1. The name of the bioengineered food and the crop from which it is derived.

2. A description of the various applications or uses of the bioengineered food, including animal fee, uses.

3. Information concerning the sources and functions of introduced genetic material.

4. Information on the purpose or intended technical effect of the modification, and its expected effect on the composition or characteristic properties of the food or feed.

5. Information concerning the identity and function of the introduced genetic material and of expressioned products encoded by the introduced genetic material, including an estimate of the concentration of any expression product in the bioengineered crop or food derived thereof.

6. Information comparing the composition or characteristics of the

bioengineered food to that of food derived from the parental variety or other commonly consumed varieties.

7. Information relating to the identity and levels of toxicants that occur naturally in the food.

8. A discussion of the available information that addresses whether the potential for the bioengineered food to induce an allergic response has been altered by the genetic modification.

9. Any other information relevant to the safety and nutritional assessment of the bioengineered food.

Source: U.S. Food and Drug Administration Office of Premarket Approval, "Guidance on Consultation Procedures: Foods Derived from New Plant Varieties" (U.S. Food and Drug Administration, October 1997).

DOCUMENT 29: Henry L. Miller, "A Need to Reinvent Biotechnology Regulation at the EPA" (1994)

This document describes and critiques the Environmental Protection Agency's method of testing the safety of agricultural products that are genetically engineered and have pesticidal properties. The focus of the author's critique is that the policies focus on the process by which these products are generated rather than determine whether the product poses a risk. The following document illustrates the problems many critics find with the EPA's method of evaluation.

* * *

The Environmental Protection Agency (EPA) has just announced final or preliminary policies for regulating various products that are genetically engineered and that have pesticidal properties (1, 2). These represent the culmination of 10 years of effort to come to regulatory terms with the advent of recombinant DNA (rDNA) technology: a set of methods that have allowed researchers and plant breeders to control and transfer more precisely than ever before the genetic traits they find desirable. The intervening years and the stellar advances in the comprehension of the scientific bases of risk that have occurred during that period have allowed ample opportunity for scientific principles to drive regulatory policy (3). Yet, even in rules and policies announced in 1994, the EPA has cast into federal regulatory policy an anachronistic approach that targets the techniques used to create these organisms—that is, rDNA methods—rather than high-risk organisms or experiments likely to pose significant risk to public health or the environment. Although the technology has become evermore sophisticated, and although numerous experiments—including risk assessment experiments—continue to demonstrate that rDNA manipulation, per se, does not confer enhanced risk (4), the EPA has steadfastly discounted these findings. Instead, the EPA has held fast to its course of singling out these organisms for special consideration—consideration that turns into burdensome and unnecessary regulatory reviews (3, 5).

Still, scientifically defensible and viable policy alternatives do exist. They have been implemented by government agencies such as the National Institutes of Health (NIH) and Centers for Disease Control (CDC) (6) and proposed by the National Research Council (7) and others (8).

Flawed Policies: Regulating Processes Instead of Risks

The EPA has a lengthy history of policy formulation based on considerations other than scientific predictions or measures of risk related to environmental protection. For example, in the late 1980s, in response to a widespread media campaign waged primarily by the Natural Resources Defense Council, the EPA pressured apple growers to abandon the use of the plant growth regulator Alar, an agricultural chemical that permits apples to ripen uniformly and increases yield. EPA's capitulation to environmentalists' demands conflicted with the agency's own scientific findings (9, 10). Environmentalists' demands appear likewise to have influenced the EPA's approach to regulating products of the new biotechnology.

The EPA's final rule for "genetically engineered" microbial biocontrol agents under the Federal Insecticide, Fungicide And Rodenticide Act (FIFRA, the "pesticide statute") was published on 1 September 1994 (1). While this regulation represented an opportunity for scientific principles to drive regulatory policy, the EPA, with the active collaboration of other parts of the administration, adopted a highly centralized, intrusive approach that once again targets techniques rather than high-risk organisms, environmental results, or outcomes.

Plants as Pesticides

The EPA has begun a process that would require the regulatory review of a whole category of products that until now have been perceived as requiring no regulation at all: whole plants genetically modified (with rDNA techniques) for enhanced pest resistance (2).

Plant varieties have long been selected by nature and humans for improved resistance or tolerance to external factors that inhibit their survival and productivity. These factors include insects, disease organisms, herbicides, and environmental stresses. All plants contain resistance traits, otherwise they would not survive. Thus, the issue is not one of the presence or absence of pesticidal properties, but one of degree. Moreover, there is no evidence to suggest that the degree of pest resistance is correlated with risk to the environment.

Genetic improvement of crops for pest resistance is likely to be less environmentally hazardous and more socially acceptable than the manufacturing and spraying of chemical pesticides. Plant breeders, farmers, and consumers possess extensive experience with crops and foods that have been genetically modified for pest resistance. In recent decades, so-called "alien" genes have been transferred widely across natural breeding boundaries by chromosome substitution or by embryo rescue techniques to yield commonly available food plants including oats, rice, black currants, pumpkins, potatoes, tomatoes, wheat, and corn (20). Most often, the desired improvement has been pest resistance, such as resistance to nematode and bacterial canker resistance in tomato, or tolate blight and leaf roll in potato (20). These are, in fact, "genetically engineered" (although not rDNA-manipulated) plants, but they are not sequestered in laboratories or test plots by regulatory mandates. Their produce is commonly available at the local supermarket or farm stand.

The EPA's proposal to regulate plants with "pesticidal" properties introduced by means of rDNA technology (2) also stands in stark contrast to the conclusions and recommendations of groups such as the National Academy of Sciences (21), the NRC (7), and the National Biotechnology Policy Board (5). The EPA is moving deliberately toward case-by-case regulation of rDNA-manipulated pest-or disease-resistant plants and would again ignore genuine risk considerations. Thus, for example, regulated field trials would include those with wheat or corn in which a gene for chitinase or one of the other newly discovered disease-resistance genes has been added by rDNA manipulation; but they would exclude a poorly characterized hybrid between a tomato and toxic, inedible members of the "deadly night-shade" family, to which it is related. The EPA would exempt from review these hybrids created by means of traditional techniques because it views these hybrids resulting from processes similar to those that occur in nature. It is ironic that the EPA's policies discriminate specifically against the new molecular techniques just as these methods are yielding a "bumper crop of disease resistant genes" that may be "the biggest thing since the discovery of chlorophyll" (22).

The adoption by the EPA of risk-based approaches to oversight would have been a win-win proposition. The advantages would have been decreased direct governmental spending on regulation, stimulation of public and private sector R&D by removing the burden of regulatory disincentives, and reassurance to the public about the essential equivalence of new biotechnology to other more familiar techniques. As it stands, however, the EPA's technique-based approach to biotechnology regulation is likely to exert a profoundly negative effect on agricultural research and on the commercialization of biological pest management strategies. Regulatory disincentives, increasingly enshrined in final regulations, will continue to deter researchers and companies from biological control strategies that could substitute safer genetically engineered microorganisms or plants for chemical pesticides. Innovations that may not provide sufficient financial return to offset the costs of testing and registration will be especially vulnerable (28). By limiting the available technological choices, the EPA's regulatory philosophy and policies are likely to damage, rather than protect, both the environment and agricultural research.

REFERENCES AND NOTES

1. *Fed. Regist.* 59, 45600 (1994).

2. Report of the Joint EPA FIFRA Scientific Advisory Pane and Biotechnology Science Advisory Committee Subpanel on Plant Pesticides, 10 February 1994.

3. H. I. Miller, in *Biotechnology*, D. Brauer, Ed. (VCH, Weinheim, Germany, in press).

4. B. W. Falk and G. Bruening, *Science* 263, 1395 (1994).

5. National Biotechnology Policy Board Report, National Institutes of Health, Office of the Director, Bethesda, MD, 1992.

6. Anonymous, *Biosafety in Microbiological and Biomedical Laboratories* (Centers for Disease Control and National Institutes of Health Department of Health and Human Services, Government Printing Office, Washington, DC, 1988).

7. *Field Testing Genetically Modified Organisms: Framework for Decisions* (National Academy Press, Washington, DC, 1989).

8. H. I. Miller, R. H. Burris, A. K. Vidaver, N. A. Wivel, *Science* 250, 490 (1990).

9. M. Fumento, *Science Under Siege* (Morrow, New York, 1993), pp. 19–44.

10. Faced with a public relations barrage against Alar, EPA Assistant Administrator John Moore issued a statement asserting that "there is inescapable and direct correlation" between exposure to UDMH (the primary degradation product of Alar) and "the development of life-threatening tumors," and that therefore the EPA would soon propose barring Alar (EPA press release, 1 February, 1989); and "EPA targets chemical used on apples," *Washington Post*, 2 February 1989, p. A4. (However, as EPA admitted separately, there were no data to support a finding of carcinogenicity.) Moore "urged" (this term has special magisterial meaning when the message comes from a high-level federal regulator) farmers who were using Alar to stop ("Hazard reported in apple chemical," *New York Times*, 2 February 1989, p. A1.

20. R. M. Goodman, H. Hauptli, A. Crossway, V. C. Knauf, *Science* 234, 48 (1987).

21. *Introduction of Recombinant DNA–Engineered Organisms into the Environment: Key Issues* (National Academy Press, Washington, DC, 1987).

22. A. S. Moffatt, *Science* 265, 1804 (1994).

28. R. J. Cook. Letter to the Environmental Protection Agency Public Response Section, Field Operations Division. 17 September 1992.

Source: Henry L. Miller, "A Need to Reinvent Biotechnology Regulation at the EPA," *Science* 266 (December 16, 1994): 1815–1818.

DOCUMENT 30: Anne Simon Moffat, "High-Tech Plants Promise a Bumper Crop of New Products" (1992)

This document presents an overview of actual and potential developments in plant genetics and biotechnology. The author presents a variety of international efforts directed to both improving plants and developing plants through various genetic engineering techniques to function as biological factories that produce elements other than those natural to them. The following document illustrates how excited the scientific community is over future use of this technology.

* * *

Agriculture is one of the all too rare U.S. industries that can hold their own in the toughest international competition. Indeed if anything, it's been too successful: Year after year it yields huge surpluses of crops that end up in storage—at great cost to the public, which ultimately has to pay both for the storage and for the government price supports that maintain farmers' incomes in the face of the production glut. Might plant biotechnology come to the rescue?

Within the past 2 to 3 years, researchers in several labs around the world have genetically engineered plants, including such common crop plants as potato and tobacco, to manufacture a wide variety of materials, including human proteins such as albumin and interferon, alpha-amylase, a bacterial enzyme that is widely used in the food processing industry, and natural polymers, including a type of polyester. The work may make it possible to divert agricultural production from surplus crops, such as feed corn, and from those of questionable social value, such as smoking tobacco, to more useful commodities, says Charles Hess, former assistant secretary for science and education at the U.S. Department of Agriculture (USDA) and now a faculty member at the University of California, Davis. "Plants have a great deal of flexibility and the potential for vast productivity," adds plant biochemist Chris Somerville of Michigan State University in East Lansing.

Using plants to manufacture new products could have "multiple societal benefits," Hess predicts. Besides reducing the need for agricultural subsidies and allowing the farmers of the southeastern United States to wean themselves from their dependence on the cigarette industry without disrupting local economies, farmlands might become renewable production factories for oils and other products that now come from a non-renewable resource, petroleum. And while it may be a decade or more before large-scale manufacture of drugs and other products in plants becomes a reality, the idea is being taken sufficiently seriously that in March of this year USDA Secretary Edward Madigan appointed a commission, known as the "Alternative Agriculture Research and Commercialization Board," to explore novel industrial uses for farm and forest products.

The researchers doing this work point out that it is just beginning, and a key obstacle to further development of the productive potential of plants is ignorance of the details of plant metabolism. Nevertheless, enough scientific obstacles have already been toppled to give real hope that plants may soon be grown to provide much more than the basics of food and fiber. So, instead of producing surpluses of a few foodstuffs, U.S. agriculture might be redirected to yield useful quantities of a broader variety of goods.

Source: Anne Simon Moffat, "High-Tech Plants Promise a Bumper Crop of New Products," *Science* 256 (May 8, 1992): 770–771.

DOCUMENT 31: David L. Wheeler, "The Search for More Productive Rice" (1995)

> Rice is one of the most important food crops in the world. This document describes the work of Susan R. McCouch, who is using modern genetics to help develop new strains of rice. Not only does she focus on yield and hardiness; taste and ease of cooking are also of importance. In addition to finding ways to produce plants with better yields, she is also interested in developing better and more rapid means of distribution of this information so that more food will be available faster.

* * *

Agricultural scientists working to improve crop yields often have tried to come up with the perfect plant, and then distributed the results to farmers around the world.

Susan R. McCouch, an assistant professor of plant breeding at Cornell University has a different approach. She starts with plants that farmers like, and she randomly stirs in some new genes from wild species. Then she picks out the best-performing plants and uses the latest tools of molecular biology to find the genes that were responsible for the improved yields.

This approach, she thinks, could create regular cycles of crop improvement and help to feed the ever-expanding world population. "Our strategy turns classical plant breeding on its head."

Ms. McCouch is starting with rice, arguably the world's most important crop. Rice is a major part of the meals of half the world's people, particularly those who live in Asia.

EFFORTS HAVE STAGNATED

Lately, however, the efforts of agricultural scientists to increase rice yields have stagnated. In the past, plant breeders worked exclusively with what they call "elite" varieties of rice, already under cultivation. They didn't want to cross those varieties with wild plants, because they had to rely on traits they could see, not on genes they couldn't find. Besides, judging by appearance, the less-productive wild plants didn't seem to have much to offer and when they were combined with cultivated varieties, the results were unpredictable.

But Ms. McCouch says she can tease out some yield-enhancing genes from plants that perform poorly in the field. She and collaborators in Colombia, China, Indonesia, Korea, and the Philippines hope that their approach will create many varieties of rice, each one thriving in a particular climate and suitable to local taste. "We are starting with local varieties and ending with local varieties," she says. "They won't look or act or behave all the same."

Modern genetics began in the mid-19th century, with Gregor Mendel's experiments with garden peas. By the late 1980s and early '90s, research administrators were accusing plant scientists of lagging behind other researchers in adopting the newest techniques of manipulating genes. Now, however, plant scientists have gained lost ground, constructing molecular maps of plants ranging from loblolly pine to sugar cane. Ms. McCouch has spent much of her career making genetic maps of rice.

2 FORMS OF SELECTION

Her applied approach to rice genetics builds on the success of Steven D. Tanksley, the Cornell scientist in whose laboratory she trained. Mr. Tanksley has grown exceptionally large tomatoes by combining a small-fruited wild species with a cultivated one. "Even in a tiny tomato that a breeder would never select for bigness, there are genes for large size," says Gary H. Toenniessen, deputy director of agricultural sciences at the Rockefeller Foundation, which has supported the work of both Mr. Tanksley and Ms. McCouch.

The reason that many agriculturally valuable genes lie hidden in wild species, plant scientists say, is that crop plants have evolved under two distinct forms of selection. In the wild, nature "picks" the plants that are good at surviving. On the farm and in university greenhouses, breeders select plants that are good at producing food.

After roughly 10,000 years of agriculture, Ms. McCouch says, farmers have ended up with cultivated species that contain only a small part of the available pool of plant genes. Yield-enhancing genes lie concealed in some wild relatives of crop plants because the plants may not need to activate them as much or as often in wild environments.

Ms. McCouch combines her attempt to improve rice with her efforts to develop maps of the plant's 12 chromosomes. The maps are based on markers that serve as signposts on the chromosomes and that help direct scientists to the target genes.

Her work has contributed to an effort to map the DNA of many other members of the grass family, including the other cereals. Jeff Bennetzen, a professor of biology at Purdue University and director of the international project, says the maps will help agricultural scientists answer increasingly complicated questions. "We're talking about changing the biology of plants so we can do completely new things or do other things much better. To do that, we need people to provide tools everyone can list. Susan has been a major player in that."

In her work, Ms. McCouch is also making use of seed collections kept at the International Rice Research Institute, in Manila. Her research is still in its early stages but she hopes that eventually, she and her collaborators can duplicate the results of "Green Revolution," which began in the late 1960s. By breeding cereal grain especially wheat and rice, so that each plant produced more food and resisted disease, researchers helped avert famine in many countries.

One product of more recent agricultural research has been hybrid rice, which combines two varieties of plants to produce high yield. But hybrid rice has a key disadvantage: Its seed is too complicated for farmers to create, and the seed they harvest from plants in the field won't generate the same high yields.

As a result, hybrid-rice seed has to be distributed by a company or a government agency each growing season. Ms. McCouch wants to create rice varieties, known as inbreds, that can produce their own seed—and she wants to make these inbreds more productive than hybrid. Such an approach, she believes, would put the power of biotechnology in the hands of farmers.

COMBINING 3 SPECIES

Her effort to use molecular maps and the the seeds of wild plants to make high-yield inbreds has been under way for two years. Some preliminary results indicate that it might work.

In one set of experiments, she has combined three species: one considered to be the Asian ancestor of all cultivated rice, one that is regularly cultivated in Africa, and a wild relative of the African cultivated rice. The offspring are being tested at the Hunan Hybrid Rice Research Center, in China, where the plants have produced yields higher than that of any of their parents. Ms. McCouch believes that the new method can generate yield increases of 3 to 5 percent a year for the next 15 to 20 years.

In its simplest form, the process begins when a wild plant is used to pollinate a plant commonly used by farmers. Although rice is self-pollinating, scientists can cross rice plants by removing the stamens, or male parts of the plant, and adding pollen from another plant to the pistil or female parts.

The offspring of that combination of plants are blended again with a cultivated plant, and the scientists study the third generation. They use this two-step crossing method because they want just a few "wild" genes from the cultivated plant, so as not to create a much different plant from the one that farmers already like.

"This is something of a blind process," says Ms. McCouch. "You have to be ready to take what comes instead of starting with a vision of what you want. But sometimes nature throws out wild and wonderful things."

GRAIN SIZE AND SHAPE

In each generation of crossed, the scientists eliminate sickly, sterile, and other obviously undesirable plants. In examining the third generation, they look a little more closely at the rice to see how it performs. They study grain size and shape, the number of grains in a seed cluster, or panicle, and the number of panicles on a plant. They also check to see how the rice cooks and tastes. "You need more than just brute yield," Ms. McCouch says.

In plants that appear to be performing well, she can use her knowledge of the genetic maps of rice to find out from which parent the yield-enhancing genes came. She can determine where the genes are on the chromosomes of both the parent and the offspring. Being able to connect particular genetic markers with plant performance speeds up the selection of the desired plants, making it possible to select a plant for agricultural use in three or four generations instead of 10 or 12, as was the case with conventional plant breeding.

In addition, the use of the molecular markers makes it possible to sort out how much each gene contributes to complex traits that are influenced by many genes, she says.

Ms McCouch's work in rice genetics also is helping scientists understand the genetics of other cereals. "The simplest genome is the best model, and rice has the simplest genome," says Mr. Bennetzen at Purdue. Scientists hunting for genes in other cereals often start their search in rice and they use the location of the gene in rice chromosomes to guess at its position in their plant of interest.

Ms. McCouch is finding a new generation of molecular markers that can be sent to other scientists as data, rather than as actual chunks of DNA packaged in yeast or bacteria. The data can be sent via electronic mail. "No quarantines, no customs, no postal delays," she says. "It's great."

A scientist in Indonesia or Colombia then can use the data to fashion molecular probes that pull out the same piece of rice DNA that a scientist at Cornell has in hand.

"Susan is clearly committed to getting things out and developing third-world collaborations and strengthening third-world programs as well as her own," says Mr. Toenniessen of the Rockefeller Foundation. "That sounds straightforward—but you'd be amazed how many academics don't want to do it."

Source: David L. Wheeler, "The Search for More Productive Rice," *Chronicle of Higher Education*, December 1, 1995, A12 and A23.

DOCUMENT 32: Barnaby J. Feder, "Out of the Lab, A Revolution on the Farm" (1996)

The use of genetically engineered agricultural products has had positive effects not only for their manufacturers but also for farmers and consumers. The following selection presents some expected benefits.

* * *

Leaders in the new technology are hard at work to expand their range to other crops, including rice and wheat. Moreover, genetically engineered products still in the pipeline should generate higher revenues for farmers instead of simply giving more business to companies with new products. Coming soon will be corn and soybeans with higher oil or protein content to make them more valuable as animal feed, and colored cotton that will reduce the need for chemical dyeing.

But the spread of genetic engineering into big commodity crops is a bigger test of its potential to help feed a world that is expected to double its population over the next 40 years.

Experts say at least half of the acreage of the nation's major crops will be covered with plants harboring at least one foreign gene in the next century. It would take only slight percentage gains in yields to justify the investments.

Right now the hottest area of debate is insect-resistant crops. Those reaching the market this year are based on genes from a family of bacteria commonly referred to as Bt, a code for genes that can kill certain pests while having no effect on beneficial insects and animals.

Consequently [genetically engineered cotton seeds with Bt] may cover as many as 2.5 million acres, about 40 percent of cotton acres traditionally requiring heavy spraying. Nucoton carries a $32-an-acre licensing fee payable to Monsanto, but should help farmers avoid $50 to $60 an acre in pesticide application in insect-infested regions.

Source: Barnaby J. Feder, "Out of the Lab, A Revolution on the Farm," *New York Times*, March 3, 1996, sec. 3, pp. 1 and 11.

DOCUMENT 33: Convention on Biodiversity (1992)

This convention, signed at the "Earth Summit" in Rio de Janeiro in 1992, has provisions for the use of genetically engineered organisms and also specifies how such resources are to be shared.

* * *

Article 8. In-situ Conservation

g. Establish or maintain means to regulate, manage or control the risks associated with the use or release of living modified organisms resulting from biotechnology which are likely to have adverse environmental impacts that could affect the conservation and sustainable use of biological diversity, taking also into account the risks to human health.

Article 14. Impact Assessment and Minimizing Adverse Impacts

a. Introduce appropriate procedures requiring environmental impact assessment of its proposed projects that are likely to have significant adverse effects on biological diversity with a view to avoiding or minimizing such effects and, where appropriate, allow for public participation in such procedures.

b. Introduce appropriate arrangements to ensure that the environmental consequences of its programmes and policies that are likely to have significant adverse impacts on biological diversity are duly taken into account.

Article 15. Access to Genetic Resources

1. Recognizing the sovereign rights of States over their natural resources, the authority to determine access to genetic resources rests with the national governments and is subject to national legislation.

2. Each Contracting Party shall endeavour to create conditions to fa-

cilitate access to genetic resources for environmentally sound uses by other Contracting Parties and not to impose restrictions that run for environmentally sound uses by other Contracting Parties and not to impose restrictions that run counter to the objectives of this Convention.

5. Access to genetic resources shall be subject to prior informed consent of the Contracting Party providing such resources, unless otherwise determined by that party.

6. Each contracting Party shall endeavour to develop and carry out scientific research based on genetic resources provided by other Contracting Parties with the full participation of, and where possible in, such Contracting Parties.

Article 16. Access to and Transfer of Technology

3. Each Contracting Party shall take legislative, administrative or policy measures, as appropriate, with the aim that Contracting Parties, in particular those that are developing countries, which produce genetic resources are provided access to and transfer of technology which makes use of those resources, on mutually agreed terms, including technology protected by patents and other intellectual property rights, where necessary, through the provisions of Articles 20 and 21 and in accordance with international law.

Article 19. Handling of Biotechnology and Distribution of Its Benefits

2. Each Contracting Party shall take all practicable measures to promote and advance priority access on a fair and equitable basis by Contracting Parties, especially developing countries, to the results and benefits arising from biotechnologies based upon genetic resources provided by those Contracting Parties. Such access shall be on mutually agreed terms.

3. The Contracting Parties shall consider the need for and modalities of a protocol setting out appropriate procedures, including, in particular, advance informed agreement, in the field of the safe transfer, handling and use, of any living modified organism resulting from biotechnology that may have adverse effect on the conservation and sustainable use of biological diversity.

Source: Convention on Biodiversity, 1992.

DOCUMENT 34: Youssef M. Ibrahim, "Genetic Soybeans Alarm Europeans" (1996)

Bioengineered crops such as soybeans are now being exported from the United States. As well as representing technical breakthroughs, the crops present enormous economic opportunities for their manufacturers. This document reports how such crops are being rejected by some Europeans. Such resistance is fostered by some environmentalist groups, but it is also supported by European food scares such as mad cow disease. The document discusses the biology, the economics, and the politics of exporting genetically engineered soybeans.

* * *

As genetically engineered soybeans from the United States begin to arrive in Europe concerns about possible health risks raised by consumer groups and critics of biotechnology are prompting a boycott by some food producers and supermarket chains.

Soybeans are one of the United States' biggest farm exports to Europe—oil from the crushed beans is used in a wide variety of grocery items, including margarine, cake and chocolate, and soymeal is fed to livestock and poultry. The bioengineered beans, developed by the St. Louis based chemical producer Monsanto to yield larger harvests at lower costs, have been approved by both the Food and Drug Administration in the United States and the European Union.

But by the time the first shipment of American soybeans that included the bioengineered bean docked in Hamburg, Germany, today aboard the freighter *Ideal Progress*, European consumer groups had made the soybean their main line of resistance against a coming wave of bioengineered crops, including corn, chicory and rapeseed. In Germany—where anxiety over tampering with the food chain has been raised to a high pitch by mad cow disease, which spread among old British cattle and, as it turned out, some diseased cows were recycled as cattle feed—packaged-food companies including Unilever and Nestle Deutschland A. G. have pledged not to use the Monsanto product.

Food industry associations in Sweden, the Netherlands and other countries are concerned that products using the beans could be challenged in court by consumer groups demanding further testing of their long-term safety. Grocery chains, including Sainsbury, Tesco and Safeway P.L.C. in Britain are demanding that American producers keep the Monsanto beans separate and labeled as altered.

In the European Union, environmental campaigners are to try next week to reverse its approval of the soybeans and to block future shipments of similarly engineered crops unless they are labeled, Hiltrud Breyer, a European Parliament Member, said in an interview the other day.

Seed companies began marketing the soybean only this year, and the variety is expected to constitute less than 2 percent of the United States' production of 59 million tons.

But Greenpeace, which sent out a barge today to interfere with the docking of the Ideal Progress, sees that trickle as the opening for a flood of bioengineered crops.

"This will be the first time genetically engineered ingredients have appeared in a wide range of every-day supermarket products," Michelle Sheather, a campaigner for Greenpeace, said in a telephone interview from headquarters in Amsterdam. "If the food multinationals have their way, millions of tons of engineered crops will be grown."

Food companies, indeed, have invested enormous amounts of money in biotechnology research to develop new varieties of crops, from corn to tomatoes, with higher yields or lower costs.

Altering genes by transplanting them from different species instead of crossbreeding related varieties, bioengineers have already made corn and potatoes more resistant to insect pests. In coming years, exports of these products are expected to total billions.

With the stakes so high, some American exporters say that the opposition in Europe in recent weeks has already caused soybean exports to the Continent to drop 10 percent. The United States Department of Agriculture is skeptical, attributing the drop in higher soybean prices at a time when European grain production is greater than usual. In any case, officials say, soybeans that cannot be sold in Europe will find a market elsewhere.

The fight against genetically engineered food products may also be fed, in part, by recent disputes in which Europeans have accused the United States of trying to control commerce beyond its borders, such as American sanctions against countries that trade with Iran, Libya and Cuba. And there may well be resentment over the assumption that if a United States agency has deemed a product safe, that should be good enough for Europeans.

Surveys have found that 85 percent to 90 percent of European consumers support clear labeling of bioengineered products, and are less enthusiastic than American consumers about embracing what the Europeans see as novel foods.

But the large food companies here agree with Monsanto that once the public is informed of the benefits of newly developed crop varieties and what the producers insist are their low risks, there will be acceptance.

"This is not in itself a big deal, but it sets signals for things to come," said Claus Conzelmann, an assistant vice president for Nestle, in a telephone interview from company headquarters in Vesrey, Switzerland. "We are perhaps a little more realistic about it," he said. "If the consumer finally does not want it, we will draw the final conclusion. But we are convinced biological engineering will remain an important field. In our opinion, it increases the efficiency of agricultural production with positive repercussions on the environment."

Monsanto, of course, is not alone. Ciba-Geigy, DuPont, Dow Chemical, Zeneca and Hoechst have also aggressively invested in genetically engineered crops. In the United States, analysts say the compelling economics of bioengineered soybeans and the crops to follow will win out. The soybeans have "been on sale in the United States and it has not caused any problems," said William Young, of the brokerage firm Donaldson, Lufkin & Jenrette, who follows the grain industry. "I can't predict politics but economics will triumph in the end," he added. "The simple truth is that organic food is a lot more expensive."

American farmers have a say, too. And they are watching developments in Europe closely before they return to Monsanto this winter to buy more bags of expensive Roundup Ready seeds, according to Philip Paarlberg, a trade specialist at Purdue University.

By the time next year's planting starts in May and June, he said, "We will be able to assess the damage properly."

Source: Youssef M. Ibrahim, "Genetic Soybeans Alarm Europeans," *New York Times*, November 7, 1996, D1 and D24.

DOCUMENT 35: Jane Rissler, *Perils amidst the Promise: Ecological Risks of Transgenic Crops in a Global Market* **(1993)**

This executive summary presents the recommendations of the Union of Concerned Scientists for evaluating the introduction of bioengineered plants into various ecosystems. The document is concerned about possible implications of such crops because the situation is novel and scientists have insufficient data available on which to make judgments.

* * *

Although the number of [genetically engineered] products at the brink of commercialization is growing, many important issues surrounding commercialization of transgenic crops are still unsettled. Among these are the impacts of such products on the sufficiency and viability of world agriculture and the question of environmental risk. For the former, there is increasing concern about the destructive effects of food production systems on the resources on which an exploding world population will depend in the twenty-first century. Many see the solution to these problems in a fundamental reorientation of agriculture toward sustainable practices. These practices will evolve from a systems based approach to the perpetual problems of yield, pest control, and soil conservation. The important question for the future is whether engineered crops will aid or retard the global transition to a sustainable agriculture. This question deserves searching public debate, which ideally should take place before the technology is allowed to become commercialized.

A second issue concerns the health and environmental risks entailed in the wide commercial use of transgenic crops. This issue has received some discussion in government and industry circles, but often from a narrow perspective that has down played the seriousness of the risks.

What environmental risks would large numbers of these different varieties of transgenic crops pose? The complete answer is far beyond the scope of this report. Ecological risks depend, among other things, on the nature of the crop, the characteristics of the added gene, and the agricultural locale.

Conclusions

1. Commercialization of transgenic crops poses serious environmental risks. The widespread commercialization of transgenic versions of the full spectrum of food and fiber crops poses serious environmental risks that can be considered in several broad categories. These include the possibilities that transgenic crops themselves will become weeds. Weeds—a term that will be used broadly in this report to cover plants unwanted in farms, lawns, roadsides, and unmanaged ecosystems—are billion-dollar problems. In 1991 alone, farmers and others spent over $4 billion to control weeds in the United States.

—transgenic crops will serve as a conduit through which new genes move to wild plants, which could then become weeds. Like the crops themselves that become weeds, these plants could require expensive control programs. In addition, the novel transgenes may affect wild ecosystems in ways that are difficult to evaluate.

—plants engineered to contain virus particles will facilitate the creation of new viruses. New viral pathogens that affect economically important crops could require significant control costs.

—plants engineered to express potentially toxic substances like drugs and pesticides will present risks to other organisms that are not the intended targets of the new chemicals. For example, drug-producing plants might poison birds feeding in corn fields.

2. Commercialization of transgenic crops could threaten global diversity.

Among the most challenging risks of transgenic crops is the threat that they might pose to the populations of wild plants and landraces (traditional varieties) that are centers of genetic diversity for crops—both as a result of the competition from the transgenic crops themselves and also by the transfer of the new genes in the crops into the landraces or wild relatives via pollen transfer. These centers of diversity, located primarily in the developing world are in regions that contain the greatest concentration of crop biodiversity. Together landraces and wild relatives are the richest repositories of crop genetic diversity. They are the natural reservoir for the traits needed to maintain the vitality of modern crops. The genes for important traits like disease resistance—few of which have been identified and isolated—are the natural capital on which both traditional crop breeders and genetic engineers depend.

Crop genetic diversity is already diminishing at a stunning rate as farmers around the world are persuaded to abandon the numerous land races of the past in favor of relatively few modern crop varieties. Expensive transgenic plants, which will generally have to create large markets to recoup research costs, will exacerbate that trend.

Few US crops have centers of diversity in this country. This means that transgenic crops sold and planted in the United States generally pose less of a direct threat to crop biodiversity than they do if they are planted elsewhere in the world. At the same time, the importance of centers of diversity to the continued vitality of US and world agriculture makes it imperative that the United States understand the vital role of these regions in the future of agriculture.

3. Two aspects of the risks of transgenic crops can be assessed and minimized through a scientifically sound regulatory system.

These are that transgenic crops themselves will become weeds and that novel transgenes will be transferred into wild populations. Weediness potential can be predicted, although imperfectly, on the basis of the comparison of the field behavior of the transgenic plant to its nonengineered parent. The possibility of transgene flow can be assessed primarily on the basis of information about the distribution of sexually compatible wild and weedy relatives in the region where the crop is grown. If no wild and weedy relatives are in the vicinity, for example, there is no risk of gene transfer, where relatives are found in the region, experiments can be done to determine the degree to which the crop and its relatives interbreed. If the transgenes are transferred to relatives, weediness potential experiments can be done.

4. Other aspects of the risk of transgenic plants are difficult to evaluate.

For example, the likelihood of gene flow is difficult to assess in many places outside the United States because information is unavailable on the distribution of wild plants. This is often true of countries harboring centers of diversity for crops important to world agriculture.

Also, the likelihood that virus-resistant crops will lead to the creation of new viruses is difficult to evaluate due to the lack of established methods to measure virus populations in the field, though there is adequate information to begin to develop such methods.

The long-term, cumulative risks to ecosystems of introducing large numbers of transgenes and transgenic plants are not well enough understood to allow their prediction except in the grossest sense. It is un-likely that ecosystem dynamics will be well enough understood any time in the near future to confidently predict this aspect of environmental impact. Obviously unknown risks cannot be anticipated or evaluated.

RECOMMENDATIONS

The Union of Concerned Scientists (UCS) calls on the federal govern-ment to adopt strong measures to protect against the domestic, and, to the extent possible, the global environmental risks posed by genetically engineered crops. UCS recommends the following specifications:

1. The United States should establish a strong federal program to as-sess and minimize the risks of transgenic crops before they are com-mercialized. The Program should consider the risks agriculture and wild ecosystems in the United States and elsewhere in the world and should pay particular attention to the protection of centers of diversity for im-portant food and fiber crops.

2. All transgenic crops should be evaluated for at least two aspects of ecological risk—weediness potential and gene flow—before they are ap-proved for commercialization. Effects on centers of diversity within the United States should receive special attention.

3. The federal government should develop standard protocols to as-sess the risks of creating new viruses, nontarget effects of pesticides, and the eco-toxicity of plant pharmaceuticals.

4. The US government should sponsor research that would enable a full assessment of all ecological risks of genetically engineered crops.

5. Congress should direct the National Academy of Sciences to pre-pare a report on i) the likelihood that seeds of engineered crops devel-oped in the United States will be dispersed to centers of crop diversity, and ii) the availability of floristic surveys and other information needed to assess the impacts of engineered crops released in countries harboring the centers.

6. All transgenic seeds that are exported from the United States should bear a label stating that approval of the seeds under US law carries no implication of safe use in other countries.

7. No company should be permitted to commercialize a transgenic crop in this country until a strong government program is in place that assures risk assessment and control of all transgenic crops, and gives adequate consideration to centers of crop biodiversity here and else-where in the world.

The appropriate United Nations organization should develop inter-national biosafety protocols, which are necessary to ensure that devel-oping countries, especially those harboring centers of crop genetic diversity, can protect against the risks of genetically engineered crops.

Source: Jane Rissler, Executive Summary, *Perils amidst the Promise: Ecological Risks of Transgenic Crops in a Global Market* (Cambridge, Mass.: Union of Concerned Scientists, 1993).

DOCUMENT 36: *Asgrow Seed Co. v. Winterboer* (1995)

This document from the Supreme Court of the United States is concerned with an implication of the development of new crops: May someone retain the seed from the crop and sell it to others? The issue is to determine to what extent such novel plants are protected as well as how to define various terms in the law and how to understand the intent of Congress when it drafted the law. The document hints at the types of questions that will continue to emerge in the wake of newly developed crops.

* * *

Syllabus

Petitioner Asgrow Seed Company has protected two varieties of soybean seed under the Plant Variety Protection Act of 1970 (PVPA), which extends patent-like protection to novel varieties of sexually reproduced plants (plants grown from seed). After respondent farmers planted 265 acres of Asgrow's seed and sold the entire saleable crop—enough to plant 10,000 acres—to other farmers for use as seed, Asgrow filed suit, alleging infringement under, inter alia, 7 U.S.C. §2541 (1), for selling or offering to sell the seed, and §2541 (3), for "sexually multiply[ing] the novel varieties as a step in marketing [them] (for growing purposes)." Respondents contended that they were entitled to a statutory exemption from liability under §2543, which provides in relevant part that "[e]xcept to the extent that such action may constitute an infringement under [§2541 (3)]," a farmer may "save seed . . . and use such saved seed in the production of a crop for use on his farm, or for sale as provided in this section: Provided, That" such saved seed can be sold for reproductive purposes where both buyer and seller are farmers "whose primary farming occupation is the growing of crops for sale for other than reproductive purposes." In granting Asgrow summary judgment, the District Court found that the exemption allows a farmer to save and resell to other farmers only the amount of seed the seller would need to replant his own fields. The Court of Appeals reversed, holding that §2543 permits a farmer to sell up to half of every crop he produces from PVPA-protected seed, so long as he sells the other half for food or feed.

Held: A farmer who meets the requirements set forth in §2543's proviso may sell for reproductive purposes only such seed as he has saved for the purpose of replanting his own acreage. (a) Respondents were not eligible for the §2543 exception if their planting and harvesting were conducted "as a step in marketing" under §2541 (3), for the parties do not dispute that these actions constituted "sexual multiplication" of novel varieties. Since the PVPA does not define "marketing," the term should be given its ordinary meaning. Marketing ordinarily refers to the act of holding forth property for sale, together with the activities preparatory thereto, but does not require that there be extensive promotional or merchandising activities connected with the selling.

(b) By reason of the proviso, the first sentence of §2543 allows seed that has been preserved for reproductive purposes (saved seed) to be sold for such purposes. However, the structure of the sentence is such that this authorization does not extend to saved seed that was grown for the purpose of sale (marketing) for replanting, because that would violate §2541 (3). As a practical matter, this means that only seed that has been saved by the farmer to replant his own acreage can be sold. Thus, a farmer who saves seeds to replant his acreage, but changes his plans, may sell the seeds for replanting under the proviso's terms. The statute's language stands in the way of the limitation the Court of Appeals found in the amount of seed that can be sold. 982 F. 2d 486, reversed.

JUSTICE SCALIA delivered the opinion of the Court.

The Plant Variety Protection Act of 1970, 7 U.S.C. §2321 et seq., protects owners of novel seed varieties against unauthorized sales of their seed for replanting purposes. An exemption, however, allows farmers to make some sales of protected variety seed to other farmers. This case raises the question of whether there is a limit to the quantity of protected seed that a farmer can sell under this exemption. In 1970, Congress passed the Plant Variety Protection Act (PVPA) 84 Stat. 1542, 7 U.S.C. §2321 et seq., in order to provide developers of novel plant varieties with "adequate encouragement for research, and for marketing when appropriate, to yield for the public the benefits of new varieties," §2581. The PVPA extends patent-like protection to novel varieties of sexually reproduced plants (that is, plants grown from seed) which parallels the protection afforded asexually reproduced plant varieties (that is, varieties reproduced by propagation or grafting) under chapter 15 of the Patent Act. See 35 U.S.C. §§161–164.

The developer of a novel variety obtains PVPA coverage by acquiring a certificate of protection from the Plant Variety Protection Office. See 7 U.S.C. §§2421, 2422, 2481–2483. This confers on the owner the exclusive right for 18 years to "exclude others from selling the variety, or offering it for sale, or reproducing it, or importing it, or exporting it, or using it in producing (as distinguished from developing) a hybrid or different variety therefrom." §2483.

Petitioner, Asgrow Seed Company, is the holder of PVPA certificates protecting two novel varieties of soybean seed, which it calls A1937 and A2234. Respondents, Dennis and Becky Winterboer, are Iowa farmers whose farm spans 800 acres of Clay County, in the northwest corner of the state. The Winterboers have incorporated under the name "D-Double-U Corporation" and do business under the name "DeeBee's Feed and Seed." In addition to growing crops for sale as food and livestock feed, since 1987 the Winterboers have derived a sizable portion of their income from "brown-bag" sales of their crops to other farmers to use as seed. A brown-bag sale occurs when a farmer purchases seed from a seed company, such as Asgrow, plants the seed in his own fields, harvests the crop, cleans it, and then sells the reproduced seed to other farmers (usually in nondescript brown bags) for them to plant as crop-seed on their own farms. During 1990, the Winterboers planted 265 acres of A1937 and A2234, and sold the entire saleable crop, 10,529 bushels, to others for use as seed—enough to plant 10,000 acres. The average sale price was $8.70 per bushel, compared with a then-current price of $16.20 to $16.80 per bushel to obtain varieties A1937 and A2234 directly from Asgrow.

Concerned that the Winterboers were making a business out of selling its protected seed, Asgrow sent a local farmer, Robert Ness, to the [LCP*689] Winterboer farm to make a purchase. Mr. Winterboer informed Ness that he could sell him soybean seed that was "just like" Asgrow varieties A1937 and A2234. Ness purchased 20 bags of each; a plant biologist for Asgrow tested the seeds and determined that they were indeed A1937 and A2234.

Asgrow brought suit against the Winterboers in Federal District Court for the Northern District of Iowa, seeking damages and a permanent injunction against sale of seed harvested from crops grown from A1937 and A2234. The complaint alleged infringement under 7 U.S.C. §2541 (1), for selling or offering to sell Asgrow's protected soybean varieties; under §2541 (3), for sexually multiplying Asgrow's novel varieties as a step in marketing those varieties for growing purposes, and under §2541 (6), for dispensing the novel varieties to others in a form that could be propagated without providing notice that the seeds were of a protected variety. (n.1)

To summarize: By reason [LCP*694] of its proviso the first sentence of §2543 allows seed that has been preserved for reproductive purposes ("saved seed") to be sold for such purposes. The structure of the sentence is such, however, that this authorization does not extend to saved seed that was grown for the very purpose of sale ("marketing") for replanting—because in that case, §2541 (3) would be violated, and the above-discussed exception to the exemption would apply. As a practical matter, since §2541 (1) prohibits all unauthorized transfer of title to or possession of the protected variety, this means that the only seed that can be sold under the proviso is seed that has been saved by the farmer to replant his own acreage. (n.5) (We think that limitation is also apparent from the text of the crop exemption, which permits a farm crop from saved seeds to be sold—for nonreproductive purposes—only if those saved seeds were "produced by descent *on such farm*" (emphasis added). It is in our view the proviso in §2543, and not the crop exemption, which authorizes the permitted buyers of saved seeds to sell the crops they produce.) Thus, if a farmer saves seeds to replant his acreage, but for some reason changes his plans, he may instead sell those seeds for replanting under the terms set forth in the proviso (or of course sell them for nonreproductive purposes under the crop exemption). It remains to discuss one final feature of the proviso authorizing limited sales for reproductive purposes. The proviso allows sales of saved seed for replanting purposes only between persons "whose primary farming occupation is the growing of crops for sale for other than reproductive purposes." The Federal Circuit, which rejected the proposition that the only seed sellable under the exemption is seed saved for the farmer's own replanting, sought to achieve some limitation upon the quantity of seed that can be sold for reproductive purposes by adopting a "crop-by-crop" approach to the "primary farming occupation" requirement of the proviso. "[B]uyers or sellers of brown bag seed qualify for the crop exemption," it concluded, "only if they produce a larger crop from a protected seed for consumption (or other nonreproductive purposes) than for sale as seed." 982 F. 2d, at 490. That is to say, the brown-bag seller can sell no more than half of his protected crop for seed. The words of the statute, however, stand in the way of this creative (if somewhat insubstantial) limitation. To ask what is a farmer's "primary farming occupation" is to ask what constitutes the bulk of his total farming business. Selling crops for other than reproductive purposes must constitute the preponderance of the farmer's business, not just the preponderance of his business in the [LCP*695] protected seed. There is simply no way to derive from this text the narrower focus that the Federal Circuit applied. Thus, if the quantity of seed that can be sold is not limited as we have described—by reference to the original purpose for which the seed is saved—then it is barely limited at all (i.e., limited only by the volume or worth of the selling farmer's total crop sales for other than reproductive purposes). This seems to us a most unlikely result.

We hold that a farmer who meets the requirements set forth in the proviso to §2543 may sell for reproductive purposes only such seed as he has saved for the purpose of replanting his own acreage. While the meaning of the text is by no means clear, this is in our view the only reading that comports with the statutory purpose of affording "adequate encouragement for research, and for marketing when appropriate, to yield for the public the benefits of new varieties." 7 U.S.C. §2581. Because we find the sales here were unlawful, we do not reach the second question on which we granted certiorari—whether sales authorized under §2543 remain subject to the notice requirement of §2541 (6). The judgment of the Court of Appeals for the Federal Circuit is

Reversed.

Justice Stevens, dissenting.

The key to this statutory puzzle is the meaning of the phrase, "as a step in marketing," as used in 7 U.S.C. §2541 (a)(3). If it is synonymous with "for the purpose of selling," as the Court holds, see ante, at 130 L.Ed.2d, at 691–692, then the majority's comprehensive exposition of the statute is correct. I record my dissent only because that phrase conveys a different message to me.

There must be a reason why Congress used the word "marketing" rather than the more common term "selling." Indeed, in §2541 (a)(1), contained in the same subsection of the statute as the crucial language, Congress made it an act of infringement to "sell the novel variety." Yet, in §2541 (a)(3), a mere two clauses later, Congress eschewed the word "sell" in favor of "marketing." Because Congress obviously could have prohibited sexual multiplication "as a step in selling," I presume that when it elected to prohibit sexual multiplication only "as a step in marketing (for growing purposes) the variety," Congress meant something different.

Moreover, as used in this statute, "marketing" must be narrower, not broader, than selling. The majority is correct that one dictionary meaning of "marketing" is the act of selling and all acts preparatory thereto. See ante, at 130 L.Ed.2d, at 691–692. But Congress has prohibited only one preparatory act—that of sexual multiplication—and only when it is a step in marketing. Under the majority's broad definition of "marketing," prohibiting sexual multiplication "as a step in marketing" can be no broader than prohibiting sexual multiplication "as a step in selling," because all steps in marketing are, ultimately, steps in selling. If "marketing" can be no broader than "selling," and if Congress did [LCP*696] not intend the two terms to be coextensive, then "marketing" must encompass something less than all "selling."

The statute as a whole—and as interpreted by the Court of Appeals—indicates that Congress intended to preserve the farmer's right to engage in so-called "brown-bag sales" of seed to neighboring farmers. Congress limited that right by the express requirement that such sales may not constitute the "primary farming occupation" of either the buyer or the seller. Moreover §2541 (a)(3) makes it abundantly clear that the unauthorized participation in "marketing" of protected varieties is taboo. If one interprets "marketing" to refer to a sub-category of selling activities, namely merchandising through farm cooperatives, wholesalers, retailers, or other commercial distributors, the entire statute seems to make sense. I think Congress wanted to allow any ordinary brown-bag sale from one farmer to another, but, as the Court of Appeals concluded, it did not want to permit farmers to compete with seed manufacturers on their own ground, through "extensive or coordinated selling activities, such as advertising, using an intervening sales representative, or similar extended merchandising or retail activities." 982 F.2d 486, 492 (CA Fed. 1992).

This reading of the statute is consistent with our time-honored practice of viewing restraints on the alienation of property with disfavor. See, e.g., *Sexton v. Wheaton*, 8 Wheat. 229, 242 (1823) (opinion of Marshall, C. J.). The seed at issue is part of a crop planted and harvested by a farmer on his own property. Generally the owner of personal property— even a patented or copyrighted article—is free to dispose of that property as he sees fit. See, e.g., *United States v. Univis Lens Co.*, 316 U.S. 241 250–252 (1942), *Bobbs-Merrill Co. v. Straus*, 210 U.S. 339, 350–351 (1908). A statutory restraint on this basic freedom should be expressed clearly and unambiguously. Cf. *Deepsouth Packing Co. v. Laitram Corp.*, 406 U.S. 518, 530–531 (1972). As the majority recognizes, the meaning of this statute is "by no means clear." Ante, at 130 L.Ed.2d, at 695. Accordingly, both because I am persuaded that the Court of Appeals correctly interpreted the intent of Congress, and because doubts should be resolved against purported restraints on freedom, I would affirm the judgment below.

Source: Asgrow Seed Co. v. Winterboer 115 S.Ct. 788, 130 L.Ed.2d 682 (1995).

Part IV

The Human Genome Project

The Human Genome Project, the constructing of a map of the whole human genome, is a multiyear, multimillion-dollar project. So vast and significant are its implications that it has often been called the Manhattan Project of biology. The documents in the following section give background to the beginning of the project, discuss its biological implications as well as problems resulting from this knowledge, and suggest religious and philosophical implications of the uses of information from the project.

DOCUMENT 37: Helen Donis-Keller and others, "A Genetic Linkage Map of the Human Genome" (1987)

The development of the Human Genome Project—the mapping of the human genome, all the genetic information contained on the human chromosome—involved many different steps. The first step was evolving in the 1980s when methods were developed to clone large fragments of DNA. By 1985, groups around the world had begun to work to find chromosome markers for linkage analysis. By 1987, the Donis-Keller group had published a linkage map of the human genome.

* * *

The notion of a genetic map dates back to 1911, when Sturtevant, while an undergraduate in T. H. Morgan's lab, realized that linkage information could be used to determine the relative position of genes along a chromosome, and at once produced the first genetic map, comprising five sex-linked loci in Drosophilia.... Over the next 75 years, complete genetic linkage maps proved to be essential tools for studying the properties of mutations. Genetic markers gained new importance with the advent of recombinant DNA, since cloned markers provide starting points for cloning closely linked genes by chromosomal walking. Unfortunately, the construction of complete genetic linkage maps has traditionally required the isolation of hundreds of single-gene mutations with easily scored phenotypes, followed by extensive interbreeding of mutant stock to ascertain the map position of the mutations.

The availability of a complete linkage map of the human genome would greatly amplify the power of this approach to human molecular genetics. With such a map, (1) the chromosomal location of newly discovered linkages could be known at once; (2) several nearby starting points would be available for efforts to clone disease genes; (3) prenatal or presymptomatic diagnosis of individuals at risk would become more accurate, through the use of markers flanking the disease gene, and more widely available, since most families would likely be informative for at least some of the markers near the disease; (4) the search for disease gene would become more efficient ... ; (5) it would become possible to map heterogeneous genetic disorders and rare recessive diseases as well as to test whether there is a genetic basis for disorders whose inheritance is currently unclear; and (6) one could begin to study the nature of recombination in humans.

Source: Helen Donis-Keller and others, "A Genetic Linkage Map of the Human Genome," Cell 51 (October 23, 1987): 319–337.

DOCUMENT 38: Renato Dulbecco, "A Turning Point in Cancer Research: Sequencing the Human Genome" (1986)

In 1986 Renato Dulbecco, president of the Salk Institute, published a short article in *Science* arguing that sequencing the human genome would be beneficial to cancer research.

* * *

We are at a turning point in the study of tumor virology and cancer in general. If we wish to learn more about cancer, we must now concentrate on the cellular genome. We are back to where cancer research started, but the situation is drastically different because we have new knowledge and critical tools, such as DNA cloning. We have two options: either to try to discover the genes important in malignancy by a piecemeal approach, or to sequence the whole genome of a selected animal species. . . . I think it will be far more useful to begin by sequencing the cellular genome. The sequence will make it possible to prepare probes for all the genes and to classify them for their expression in various cell types at the level of individual cells by means of cytological hybridization.

An effort of this kind could not be undertaken by any single group: it would have to be a national effort. Its significance would be comparable to that of the effort that led to the conquest of space, and should be carried out with the same spirit. Even more appealing would be to make it an international undertaking, because the sequence of the human DNA is the reality of our species, and everything that happens in the world depends on those sequences.

Source: Renato Dulbecco, "A Turning Point in Cancer Research: Sequencing the Human Genome," *Science* 231 (March 7, 1986): 1055–1056.

DOCUMENT 39: Report of the National Research Council Committee on Mapping and Sequencing the Human Genome (1988)

The National Research Council (NRC) of the National Academy of Sciences appointed a committee that in 1988 issued a report that argued for a broader human genome project. Presented here are excerpts from the report.

* * *

Mapping

Full-scale mapping (not gene by gene), both genetic linkage and physical, should begin immediately.

—Because the technology needed for genetic linkage mapping with RFLPs is more advanced than that for physical mapping, an immediate emphasis should be placed on completing the genetic linkage map.

—All types of maps . . . need to be coordinated as part of a human genome project. Encourage researchers' natural tendency to construct detailed maps of chromosomal regions of particular interest.

—The committee specifically recommends against a centrally imposed plan to proceed from lower to higher resolution as is implicit, for example, in proposals to complete the entire physical map before initiating pilot sequencing projects.

—Development and refinement of techniques should be emphasized early in the mapping part of the project.

Information Handling

—All human map data should be accessible from a single data base.

—Encourage the activities of those individuals who combine skills in computer programming and biology as they will be needed to generate the DNA sequence search routines of greatest utility to the biological community.

Implementation

—Funding ought not to be provided at the expense of currently funded biological research.

—Competition between centers should be encouraged.

Funding Projection

—Estimated cost: $200 million per year, to be reached during the third year of the project.

—Estimate is based on a projected total of 1,200 individuals at $100,000 annually.

Source: National Research Council Committee on Mapping and Sequencing the Human Genome, 1988, found at http://www.nhgri.nih.gov/HGP/Historical, nrc.html.

DOCUMENT 40: U.S. Department of Energy, "Understanding Our Genetic Inheritance" (1996)

This U.S. Department of Energy report builds on past work and specifies the goals of the Human Genome Project for its first five years. Excerpts are provided.

* * *

The Human Genome Initiative is a world wide research effort that has the goal of analyzing the structure of human DNA and determining the location of all human genes. In parallel with this effort, the DNA of a set of model organisms will be studied to provide the comparative information necessary for understanding the functioning of the human genome. The information generated by the human genome project is expected to be the source book for biomedical science in the 21st century. It will have a profound impact on and expedite progress in a variety of biological fields, including those such as developmental biology and neurobiology, where scientists are just beginning to understand the underlying molecular mechanisms. The analysis and interpretation of the information will occupy scientists for many years to come. Thus, the maximal benefit of the human genome project will only be achieved if it is surrounded by research efforts that are focused on understanding and taking advantage of the human genetic information.

The human genome project is expected to immensely benefit medical science. It will help us to understand and eventually treat many of the more than 4000 genetic diseases that afflict mankind, as well as the many multi-factorial diseases in which genetic predisposition plays an important role. New technologies emanating from the genome project will also find application in other fields such as agriculture and the environmental sciences. They will be valuable for assessing the effects of radiation and other environmental factors on human genetic material.

It is anticipated that the private sector will derive great benefit from the trained manpower, the data, and the techniques developed by the human genome program and will develop many useful applications based on the new knowledge that is produced. Within a few years, DNA sequence information will undoubtedly be a major tool in most areas of basic and applied biological research.

As refined through the discussion over the last half of the 1980's and defined in the NCR report, the Human Genome Initiative has several interrelated goals:

—construction of a high-resolution genetic map of the human genome;

—production of a variety of physical maps of all human chromosomes and of the DNA of selected model organisms, with emphasis on maps that make the DNA accessible to investigators for further analysis;

—determination of the complete sequence of human DNA and of the DNA of selected model organisms;

—development of capabilities for collecting, storing, distributing, and analyzing the data produced;

—creation of appropriate technologies necessary to achieve these objectives.

Source: U.S. Department of Energy, "Understanding Our Genetic Inheritance: The U.S. Human Genome Project. The First Five Years: Fiscal Years 1991–1995." NIH Publication, 1996, No. 90–1590.

DOCUMENT 41: Carol A. Tauer, "The Human Significance of the Genome Project" (1992)

Of particular concern when examining the Human Genome Project is how new genetic information will affect our idea of what is normal and specifically how one might draw a distinction between what is normal and what is abnormal. This issue has important medical and social implications, particularly with respect to insurance. Also, increased knowledge in genetics means increased choices in reproduction because we will have a much more complete genetic profile of the individual. Do families have an obligation to share such information with members of the extended family? Can family members demand access to this information from other family members? Issues such as these will continue to be important elements of the new genetics. The following document presents the views of a philosopher at St. Catherine's University (St. Paul, Minnesota) on these issues. She takes a look at the worthiness of genetic research in its own right.

* * *

When Barbara Walters presented a television special on our increasing knowledge of human genetics, she called her program "The Perfect Baby." The choice of title was understandable; if she had announced a program on "The Human Genome Project," she might have lost most of her viewers. By explaining the genome project in the context of the search for genetic health and wholeness, Walters highlighted the connections between science and human life that are crucial to all of us.

Because scientific advances affect our lives so deeply, they raise philosophical, religious and moral questions. Progress in human genetics raises particularly troubling questions: What does it mean to be human? Are there things about ourselves it would be better not to know? How should we use our expanding genetic knowledge and power?

Why the Human Genome Project?

The human genome project is a 15-year, international program to map and sequence all the genes on the human chromosomes. Technically, mapping is the process of locating the position of the genes on the chromosomes, while sequencing identifies the nucleotides, the 3 billion base pairs which comprise the genes of one human being.(1)

In order to gain support by the U.S. Congress for this immense program of basic research, scientists have had to tout the benefits the project will produce. Unfortunately, an examination of some testimony raises serious questions about the ability of scientists to be objective when their own pet projects and scientific dreams hang in the balance. Consider the following examples from two enthusiastic supporters of full funding for the genome project.

James B. Watson, former director of the National Center for Human Genome Research, writing with Robert Cook-Deegan, recognizes that diseases like cancer and heart disease are too complex to be explained by the presence of a gene, yet claims (without explanation) "it is nonetheless clear that studying genetics is a fast track to understanding them."(5) Watson and Cook-Deegan acknowledge that diseases rampant in the developing world are most efficiently controlled through improved nutrition, housing, sanitation and relief of poverty. Yet they devote extensive discussion to the advantages of DNA-based tests over other clinical diagnosis in developing countries and attach significance to a totally hypothetical situation: "If developing countries succeed in diminishing the impact of infectious disease, their health problems will come to more closely resemble those found in the developed world today."(5) Apparently we are to conclude that in the long run, what the poorest countries really need is the genome project and the genetic information it will provide.

Watson and Cook-Deegan make inferences far beyond the scientific data they provide. Daniel Koshland, editor of *Science*, makes similar leaps in his advocacy of the human genome project as the cure for social ills. Although he recognizes the effect of environmental factors on crime, poverty and homelessness, he believes it would be unethical not to put our trust in the genome project; we would incur "the immorality of omission—the failure to apply a great new technology to aid the poor, the infirm and the underprivileged. In response to a question about using available funds to help the homeless directly, he said, "What [you] don't realize is that the homeless are impaired. . . . Indeed, no group will benefit more from the application of human genetics."(7)

Comments like these, coming from people who are perceived as "scientific" in their outlook, do not help public understanding of what the genome project realistically will accomplish in the foreseeable future. It is unfortunate that some scientists believe they need to use exaggeration and manipulation in order to achieve their goals. For the goal of mapping and sequencing the human genome is a worthy aim in many respects.

The project is achievable within a reasonable length of time, perhaps 15 years. The research does not require costly equipment, as would be the case in physics; hence it can be carried out by many people in decentralized locations. It involves scientific cooperation up to the international level. Partial results have value, even before the project as a whole is completed. It will stimulate advances in other areas, for example, in information science, which will have to provide new tools for information storage, retrieval, classification and coding. The human gene sequences discovered will be available as tools to all scientists, rather than controlled by individual countries or companies.(8) As tools, these sequences may be used to study gene function, to develop methods of screening for diseases, and to investigate genetic therapies and other new treatments. A realistic appraisal of the benefits of the genome project gives substantial support to requests for significant funding.

Philosophical and Religious Concerns

Viewed broadly, the human genome project has two dimensions: gaining knowledge through scientific research, and applying this knowledge through the techniques of genetic engineering. The philosophical and religious questions raised most frequently are about the applications, or genetic engineering. Few commentators oppose basic biological research, though they may advise scientists to proceed cautiously because of potential practical applications.

Three commonly expressed worries about genetic engineering are the concern that it involves "playing God"; the belief that it is "interfering with nature"; and the fear that some malevolent person(s) will cause terrible harms (the "Hitler" scenario). While I appreciate the seriousness and depth of these concerns, I believe they are not the real philosophical and religious problems raised by the human genome project.

"Playing God" is a phrase often used to describe interventions into the processes of human life and death. (Oddly, it is not often used in reference to environmental disruptions which probably cause more far-reaching interventions into God's creation.) We accept many such interventions with little question: Caesarean sections for high-risk births, kidney dialysis and transplants for impaired renal function, neonatal in-

tensive care technology for premature newborns and surgical removal
of cancerous tumors, organs and body parts. When such procedures improve the quality of life, or extend human life in a reasonable way, we
believe that God would approve, and perhaps require, their use. The use
of genetic interventions for the same purposes seems to deserve similar
assessment. There is no reason to judge genetic technologies differently
from other clinical modalities.

"Interference with nature" has a long history; the meaning of "nature"
and the norms of natural law have been discussed since classical Greece.
As we accept medical interventions into life and death, we accept numerous other human enterprises which change or overcome the course
of nature. Human civilization represents our efforts to mold nature to
human ends.

The ability to change human nature itself may appear to be a new
element introduced by genetic engineering, but this is misleading.
Through education, social institutions, family and reproductive practices,
we do many things to change ourselves. Genetic technologies may give
us the power to speed up some of these processes, and our technologies
surely might outrun our philosophical and moral deliberations. But the
ability to see more clearly the consequences of our choices could make
it possible for us to choose more consciously and with greater awareness.

Theologian Jean Porter suggests a criterion for moral assessments of
genetic "interference with nature": While the many operations that sus-
tain our life, including the transmission of information through the genetic code, are worthy of admiration and even awe, we need not refrain
from intervening in them when doing so would serve some legitimate
purpose. . . . What counts as natural, in a normative sense, is determined
by what is characteristic of human life as a whole, namely, our capacities
for intelligent self-direction.(9) In other words, human nature is primarily reflective, rational and purposeful, and we are in accord with that
nature when we act reflectively, rationally and purposefully.

Scenarios which describe the fear that malevolent persons will obtain
genetic technologies and use them to control and terrorize are quite easily dismissed. Not because human beings are incapable of such evils, but
because there are ample tools available for these purposes at the present
time, and current tools are much more easily and efficiently employed
than are genetic manipulations. Think of nuclear devices and explosives,
of toxic chemicals ready for release, of biological and germ weaponry
and of the potential for creating chaos simply through terrorist disruption of communication, transportation and energy infrastructures. It is
clear that those who are able and willing to seize power for evil purposes
have more than adequate means at their disposal already. The technical
difficulties of making genetic changes, which by all accounts will persist

for decades and even centuries, rule out the "Hitler" scenario as a re-alistic one.

By my somewhat cursory dismissal of three common concerns, I do not mean to suggest that the human genome project is philosophically unproblematic. In actuality, it raises deep philosophical and moral questions which also impinge on religious beliefs.

Our Image of Ourselves

In my view, the changes the human genome project could bring about in how we think of ourselves are more dangerous than the changes the project will make possible biologically. Our image of ourselves as human beings could be radically modified long before we are able to make significant permanent genetic alterations. Thus, the alterations we eventually choose would flow out of a revised concept of humanness. But it is the concept, the image, that has the power; hence that concept or image should be our concern.

The human genome project carries a dramatic biological metaphor: the notion that our genes are the program that determines who we are, and that when we know all the genes we will know the human being, both generically and individually. Scientists working on the genome project may express skeptical questions, as Victor McKusick did at a recent conference: When we know the last nucleotide, will we know what it is to be human?(10) But the implication is that in some sense we will. Regardless of disclaimers, the message that "You are your genes" is being communicated. And with testimony from scientists suggesting that all human ills, even homelessness, are best approached through genetic manipulation, this message is continually reinforced.

Scientific reductionism, the explanation of human life and behavior entirely through physical causes, becomes one notch narrower in the genetic metaphor. In this metaphor, human life and behavior are reduced to one specific type of physical cause: one's genetic heritage.

Those who have struggled against scientific reduction during the twentieth century, notably religious believers, have offered alternative conceptions of human nature and human life. These conceptions may be expressed in terms of spiritual dimensions or transcendent qualities. Their common element is the perception of an incorporeal essence, a crucial aspect of the human being which cannot be reduced to scientifically measurable components. The reductionism implicit in the genome project offers a substantial challenge to such views, not because of what is scientifically provable, but because of the impact of the language which describes the project.

In suggesting a reductionist view of human life, the genetic metaphor also implies a determinist interpretation of human behavior. If in some sense I am programmed or determined by my genes, then how can I be held responsible for what I do, or the kind of person I turn out to be? Such questions, if taken seriously, would challenge our customary moral and legal concepts of personal responsibility. Our approaches to praise and blame, reward and punishment, go back to Aristotle, who argued that we respond in those ways because we believe someone was able to make a choice. A person unable to exercise choice would deserve neither praise nor blame. Aristotle taught us to distinguish those who are responsible from those who are not. The determinist implication of the genetic metaphor is that no one is really responsible.

The Concept of Normality

Another concept which the human genome project leads us to rethink is the notion of normality. What should we regard as normal for a human being, and what should we attempt to correct or modify because it is deemed abnormal? How is the distinction drawn between normal and abnormal characteristics?(11) And by whom? Prospective parents, medical professionals, society, Madison Avenue advertising firms? The characteristics of interest range from purely physical endowments such as height, to multifactorial qualities such as intelligence, to character traits such as degree of aggressiveness. It is easy to see the pitfalls involved in applying concepts such as normal and abnormal to these characteristics. It may be less obvious that distinguishing health from disease, classifying them as normal or abnormal physical conditions, is also fraught with difficulties.

While the human genome project aims to identify the entire complement of human genes, its immediate practical interest lies in identifying the genes linked to abnormal conditions. A person who has an "abnormal" gene may actually be ill, that is, show symptoms of the disease associated with the gene. But there are three other possible meanings for a diagnosis of genetic abnormality.

A person could have one gene for a recessive disease such as Tay-Sachs or cystic fibrosis. This person does not have the disease but is a carrier, and could transmit it to offspring if the other parent is also a carrier. Secondly, a person's genes may indicate that he or she has a disease which has not yet expressed itself, for example, Huntington's chorea or polycystic kidney disease. Before this person becomes ill (the presymptomatic period) he or she might be described as healthy and diseased at the same time. Thirdly, one's genes may suggest a predisposition to a disease like breast or colon cancer, heart disease or Alzheimer's disease. Predisposition provides information similar to a family history; it indicates that one is more likely than average eventually to have this disease.

When complete genetic screening becomes possible, approximately 1,000 to 2,000 genes linked to abnormal conditions will be diagnosable.

A person screened for all of them would be expected to show 20 to 40 genetic abnormalities. Some of these would be false positives, or test results which are interpreted erroneously. But even the true positives could be misunderstood if they are described as abnormalities.

Ordinarily, a physical abnormality is defined in terms of an inability to function in a way that is normally expected.(12) But a person who is merely a carrier of a recessive disease does not lack any functional ability. Carrier status may affect one's children, but not one's own health or functioning. In the past, some persons who carried the gene for sickle cell anemia were mistakenly perceived as unhealthy and suffered discrimination.(13)

The person who is genetically identified as having a late-onset disease such as Huntington's chorea may also suffer social and economic consequences. Even though he or she shows no symptoms and is as functional as any healthy person, the genetic condition may be assessed as abnormal and may be a detriment to obtaining employment or insurance. Similarly, a person identified as genetically predisposed to contract cancer or heart disease may be viewed as abnormal in comparison with expectations of the populations as a whole.

Thus, we have three genetic identifications of abnormality which do not represent one's actual current ability to function at an expected level, but which could produce the same social or economic consequences as a lack of functional capacity. In fact, given certain life choices or environments, the actual illnesses may never be expressed. If the condition exists merely in one's genes, do we have a new sense of the term "abnormal"?

Immediate Ethical Problems

It is important that we contemplate the long-term effects which the genome project may have on our ways of thinking about ourselves, about our spiritual aspect and about responsibility and normality. But what are some of the more immediate problems which the project presents to us?

Even this early, a consensus on ethical problems has emerged among scientists, moral philosophers and legal scholars. Thoughtful people in a variety of fields believe that the immediate problems will be found in the area of genetic screening rather than genetic engineering.(14) Soon it will be possible to diagnose hundreds of genetic anomalies without being able to do much about them therapeutically. The genome project will bring about a widening gap between what we know how to diagnose and what we know how to treat.(15)

References

1. Victor A. McKusick, "Mapping and Sequencing the Human Genome," *New England Journal of Medicine* 320 (April 8, 1989), pp. 910–915.
5. James D. Watson and Robert M. Cook-Deegan, "The Human Genome Project

and International Health," *Journal of the American Medical Association* 263 (June 27, 1990), pp. 3322–3324.

7. Daniel Koshland, Jr., cited in lecture by Evelyn Fox Keller, "Nature, Nurture, and the Human Genome Project," College of St. Catherine, St. Paul, MN, September 24, 1990.

8. McKusick, "Mapping and Sequencing the Human Genome," p. 912; Charles R. Cantor, "Orchestrating the Human Genome Project," *Science* 248 (April 6, 1990), pp. 49–51.

9. Jean Porter, "What Is Morally Distinctive About Genetic Engineering?" *Human Gene Therapy* 1 (1990), pp. 419–424.

10. McKusick, "The Process and Promise of the Genome Project."

11. See Georges Canguilhem, *The Normal and the Pathological* (New York: Zone Books, 1989).

12. Dan W. Brock, "The Genome Project and Human Identity," conference "Legal and Ethical Issues," March 8, 1991.

13. President's Commission for the Study of Ethical Problems in Medicine, *Screening and Counseling for Genetic Conditions* (Washington, D.C.: U.S. Government Printing Office, 1983), pp. 20–22.

14. See, for example, Alexander M. Capron, "Which Ills to Bear? Reevaluating the 'Threat' of Modern Genetics," *Emory Law Journal* 39, no. 3 (Summer 1990), pp. 665–696.

15. Victor A. McKusick, "The Process and Promise of the Genome Project," Conference, "Legal and Ethical Issues," March 7, 1991.

Source: Carol A. Tauer, "The Human Significance of the Genome Project," *Midwest Medical Ethics*, summer 1992, pp. 3–12.

DOCUMENT 42: Daniel E. Koshland, Jr., "Sequences and Consequences of the Human Genome" (1989)

In an editorial in *Science*, Daniel E. Koshland, president of the American Association for the Advancement of Science, argued in favor of the genome project because of its benefits for all. How these benefits would occur, however, was not explained.

* * *

The benefits to science of the genome project are clear. Illnesses such as manic depression, Alzheimer's, schizophrenia, and heart disease are probably all multigenetic and even more difficult to unravel than cystic fibrosis. Yet these diseases are at the root of many current societal problems. The costs of mental illness, the difficult civil liberties problems they cause, the pain to the individual, all cry out for an early solution that involves prevention, not caretaking. To continue the current warehousing or neglect of these people, many of whom are in the ranks of the homeless, is the equivalent of providing iron lungs to polio victims at the expense of working on a vaccine.

Sequencing the human genome puts us on the threshold of great new benefits and some real but avoidable risks. There are immoralities of commission that we must avoid. But there is also the immorality of omission—the failure to apply a great new technology to aid the poor, the infirm, and the underprivileged. We must step boldly and confidently across the threshold.

Source: Daniel E. Koshland, Jr., "Sequences and Consequences of the Human Genome," *Science* 246 (October 13, 1989): 189.

DOCUMENT 43: *Tarasoff v. Regents of the University of California* (1976)

In ethical discussions a high premium is placed on confidentiality. The primary reason is to protect the interests of the patient, to ensure that patients feel free to seek therapy, and so to ensure they disclose all information relevant to their treatment. The document cited below shows one way in which the breaking of confidentiality was resolved. How this could be applied to genetic information is yet to be resolved.

* * *

We realize that the open and confidential character of psychotherapeutic dialogue encourages patients to express threats of violence, few of which are ever executed. Certainly a therapist should not be encouraged routinely to reveal such threats; such disclosures could seriously disrupt the patient's relationship with his therapist and with the persons threatened. On the contrary, the therapist's obligations to his patient require that he not disclose a confidence unless such disclosure is necessary to avert danger to others, and even then that he do so discreetly, and in a fashion that would preserve the privacy of his patient to the fullest extent compatible with the prevention of the threatened danger.

If the exercise of reasonable care to protect the threatened victim requires the therapist to warn the endangered party or those who can reasonably be expected to notify him, we see no sufficient societal interest that would protect and justify concealment. The containment of such risks lies in the public interest.

Source: Tarasoff v. Regents of the University of California, California Reporter 14 (July 1, 1976).

DOCUMENT 44: Mark A. Rothstein, "Genetic Screening in Employment: Some Legal, Ethical, and Societal Issues" (1990)

Mark Rothstein's comments in the following document echo some of these fears, in which he points out that responsibility lies with those who can choose to use this research in either good ways or bad ways.

* * *

The use of increased medical screening including genetic screening for employers may have the effect of denying health care to those most in need of it. It also could contribute to the other social ills caused by long-term unemployment, such as increases in mental illness, alcoholism, suicide and domestic violence. Taken to its extreme, genetic screening could create a perpetual genetic underclass of individuals, excluded from the mainstream of society.

Source: Mark A. Rothstein, "Genetic Screening in Employment: Some Legal, Ethical, and Societal Issues," *International Journal of Bioethics* 1, no. 4 (1990): 239–240.

DOCUMENT 45: National Institutes of Health Workshop Statement: "Reproductive Genetic Testing: Impact on Women" (1991)

How genetic testing is implemented is of significance both to individuals and to society. The following document presents excerpts from a National Institutes of Health workshop on this issue.

* * *

Reproductive genetic testing, counselling and other genetic services can be valuable components in the reproductive health care of women and their families; they can also have negative effects on individuals, families, and on communities. These services have the potential to increase knowledge about possible pregnancy outcomes that may occur if a woman decided to reproduce; to provide reassurance during pregnancy; to enhance the developing relationship between the woman, her expected child and others; to allow a woman an opportunity to choose whether on not to continue a pregnancy in which the expected child has a birth defect or a genetic disorder, and if continuing, to facilitate pre-natal or early infant therapy for her expected child, when possible; and to prepare for bearing and rearing a child with a disability. Conversely, these services have the potential to increase anxiety; to place excessive responsibility, blame and guilt on a woman for her pregnancy outcome; to interfere with maternal-infant bonding, and to disrupt relationships between a woman, family members, and her community.

1. Reproductive genetic services should not be used to pursue 'eugenic' goals, but should be aimed at increasing individuals' control over their own reproductive lives. Therefore, new strategies need to be developed to evaluate the successes of such services.

2. Reproductive genetic services should be meticulously voluntary.

3. Reproductive genetic services should be value-sensitive.

4. Standards of care for reproductive genetic services should emphasize genetics information, education and counseling rather than testing procedure alone.

5. Social, legal and economic constraints on reproductive genetic services should be removed.

6. Increasing attention focused on the development and utilization of reproductive genetic testing services may further stigmatize individuals affected by a particular disorder or disability.

Source: NIH Workshop Statement: "Reproductive Genetic Testing: Impact on Women," November 21, 23, 1991; full text found in *Fetal Diagnostic Therapies* 8 (1993) (suppl. 1): 6–9.

DOCUMENT 46: National Institutes of Health–Department of Energy, Ethics, Law, and Social Issues Working Group and National Action Plan (1995)

A critical problem raised by the Human Genome Project is the possibility of the use of genetic information in a discriminatory way either in employment or in insurance applications. The following documents provide various approaches to this concern.

* * *

The ELSI Working Group and the National Action Plan for Breast Cancer (NAPBC) developed a series of recommendations and definitions for state and federal policy makers to protect against genetic discrimination.

Recommendations

1. Insurance providers should be prohibited from using genetic information, or an individual's request for genetic services, to deny or limit any coverage or establish eligibility, continuation, enrollment, or contribution requirements.

2. Insurance providers should be prohibited from establishing differential rates or premium payments based on genetic information on an individual's request for genetic services.

3. Insurance providers should be prohibited from requesting or requiring collection or disclosure of genetic information.

4. Insurance providers and other holders of genetic information should be prohibited from releasing genetic information without prior written authorization of the individual. Written authorization should be required and include to whom the disclosure would be made.

Source: National Institutes of Health–Department of Energy Ethics, Law, and Social Issues Working Group and National Action Plan Breast Cancer Workshop on Genetic Discrimination and Health Insurance, Bethesda, Md.; July 19, 1995. found at http://www.nhgri.nih.gov/About_NHGRI/Der/Elsi/napbc.html.

DOCUMENT 47: The Genetic Information Nondiscrimination in Health Insurance Act of 1997 (Proposed)

SECTION 2706. PROHIBITION OF HEALTH INSURANCE
DISCRIMINATION ON THE BASIS OF GENETIC INFORMATION

(a) IN GENERAL—In the case of benefits consisting of medical care provided under a group health plan or in the case of group health insurance coverage offered by a health insurance issuer in connection with a group health plan, the plan or issuer may not deny, cancel, or refuse to renew such benefits or such coverage, or vary the premiums, term, or conditions for such benefits or such coverage, for any participant or beneficiary under the plan

(1) on the basis of genetic information; or

(2) on the basis that the participant or beneficiary has requested or received genetic services.

(b) LIMITATION ON COLLECTION AND DISCLOSURE OF INFORMATION

(1) IN GENERAL—A group health plan, or a health insurance issuer offering group health insurance coverage in connection with a group

health plan, may not request or require a participant or beneficiary (or an applicant for coverage as a participant or beneficiary) to disclose to the plan or issuer genetic information about the participant, beneficiary, or applicant.

(2) REQUIREMENT OF PRIOR AUTHORIZATION—A group health plan, or a health insurance issuer offering health insurance coverage in connection with a group health plan, may not disclose genetic information about a participant or beneficiary (or an applicant for coverage as a participant or beneficiary) without the prior written authorization of the participant, beneficiary, or applicant or of the legal representative thereof. Such authorization is required for each disclosure and shall include an identification of the person to whom the disclosure would be made.

Source: H. R. 306, Genetic Information Nondiscrimination in Health Insurance Act of 1997.

DOCUMENT 48: The Genetic Confidentiality and Nondiscrimination Act of 1997 (Proposed)

SECTION 401. DISCRIMINATION BY EMPLOYERS OR POTENTIAL EMPLOYERS

(a) IN GENERAL—An employer shall not request, require, or use the genetic information of an employee or a prospective employee for the purpose of restricting any right or benefit otherwise due or available to the employee or the prospective employee. An employer may request or require or use the genetic information of an employee for the purpose of—

(1) permitting a genetically susceptible employee to avoid occupational exposure to substances with a mutagenic or teratogenic effect; or

(2) determining a genotype that is otherwise directly related to the work and is consistent with business necessity.

Source: S. 422, Genetic Confidentiality and Nondiscrimination Act of 1997.

DOCUMENT 49: The Genetic Justice Act (Proposed) (1997)

SECTION 3. EMPLOYER PRACTICES

It shall be unlawful employment practice for an employer—

(1) to fail or refuse to hire or to discharge any individual, or otherwise,

to discriminate against any individual with respect to the compensation, terms, conditions, or privileges of employment of the individual, because of genetic information with respect to the individual, including an inquiry by the individual regarding genetic services.

(2) to limit, segregate, or classify the employees of the employer in any way that would deprive or tend to deprive any individual of employment opportunities or otherwise adversely affect the status of the individual as an employee, because of genetic information with respect to the individual, including an inquiry by the individual regarding genetic services; or

(3) to request or require the collection for the employer or disclosure to the employer of genetic information with respect to an individual unless the employer shows that—

(A) the employer made the request or requirement after making an offer of employment to the individual;

(B) the information is job-related for the position in question and consistent with business necessity; and

(C) the knowing and voluntary written consent of the individual has been obtained for the request or requirement, and the collection or disclosure.

Source: S. 1045, Genetic Justice Act (1997).

DOCUMENT 50: Nachama L. Wilker and others, "DNA Data Banking and the Public Interest" (1992)

Concerns and fears about the use of genetic information extend to the storage of and access to genetic information in DNA data banks. The following are recommendations from the Human Genetics Committee of the Council for Responsible Genetics in Cambridge, Massachusetts.

* * *

1. Genetic information gathered for prenatal, neonatal, transplantation, blood typing, identification, and other screening purposes by private or public agencies must be restricted to the specific purpose for which it was collected.

2. Until safeguards are in place to prevent misuse of genetic information or stored biological samples, data bands designed for identification purposes must not contain biological samples and should be used only with the donor's permission. When this is not possible, an impartial

judicial review should control access to this information, and a "necessity" or "public good" standard should be met before allowing any non-consensual access.

3. Data banks must be open to public scrutiny.

4. The storage of biological samples in data banks must not be routinely undertaken by any agency, except those with a medical purpose, and then only when consent, confidentiality, and security are well established.

5. Government data banks that store genetic information or biological samples should not be used for insurance-related inquiries, employment decision, or external agency review. . . . For example, the FBI should be forbidden from using the DOD's genetic data bank.

Source: Nachama L. Wilker and others, "DNA Data Banking and the Public Interest," in Paul R. Billings, ed. *DNA on Trial: Genetic Identification and Criminal Justice* (New York: Cold Spring Harbor Laboratory Press, 1992), 141–149.

DOCUMENT 51: Ted F. Peters and Robert J. Russell, "The Human Genome Project: What Questions Does It Raise for Theology and Ethics?" (1992)

The following document raises several explicitly theological or religious issues related to the Human Genome Project. Genetics might force a reconsideration of ourselves as created in the image of God. There are also implications of understanding various religious perspectives on human nature such as our capacity to love, our capacity to be creative, and our capacity to intervene in nature. Such questions provoke other questions related to the status of the family and the genetic links traditionally understood to be the basis of relationships. The main point is that genetics involves more than the study of the genes, and consideration of issues such as these is of critical importance.

* * *

We can picture the day when shortly after conception, whether in a test tube or in the mother's womb, each child can be tested, his or her genome decoded, and either abortion or therapeutic intervention begun. We can similarly imagine the day when each individual will be identifiable by his or her genetic code, a code that like our social security number will precede or follow us wherever we go. We can in addition foresee the day when someone will embark on positive eugenics, not only seeking therapy for those with existing genetic defects, but pushing on to engineer a germ line of "new and improved" versions of the human being, perhaps with higher intelligence or greater physical strength or a longer life span. In short, we can actually conceive of a Brave New World scenario arriving within a generation.

The human genome project will likely send theologians to their libraries to review and rethink our relationship to God; and it will send ethicists to their desks to think through the social, political, economic and legal implications. Let us look at some of the questions likely to arise as genetic research enters this new phase.

Theological Questions

The first question is this: What is the nature of human nature? This question is shared by scientists and theologians alike. It has already been posed by James Watson, until recently the project's national director, when he said that the goal of research is "to find out what being human is."

Theologians like to refine this question by asking just how our human nature is determined by our relation to God. Jews and Christians normally begin to answer this question by referring to the book of the Bible from which genes derive their name, Genesis, and the claim that we humans are created in God's image. We humans are the imago dei. But what does it mean to be created in God's image, especially in light of our own potential ability to alter the genetic constitution of the human being? How does "image" relate to "nature," divine image to biological nature, and creation in this image to evolution of the human genome? Before proceeding, perhaps we should remind ourselves of the ways in which the imago dei has been thought of in the past. Plato and his heirs in Western philosophy and religion sought the connection between the human and the divine in the mind: God thinks; so do we. God is rational; so, if we wish to enhance our Godlikeness, we should develop our rational potential. If we as humans can train our minds to love rational ideas—especially eternal ideas as unchanging forms we will realize a oneness with the divine that carries us beyond the veil of this temporal and temporary world. It will carry us beyond time to eternity. Human nature, according to this view, finds its connection with God through a mind that concentrates on that which is fixed and unchanging.

Others in the history of theology have sought to locate the imago dei in the heart or the will. Here, our capacity to love becomes decisive. "God is love," says the Bible in 1 John 4:16. We can love, too. Jesus makes it clear that divine love is agape, self-sacrificing love. We too can love self-sacrificingly, even if we do not often avail ourselves of the opportunity. Some have combined the love ethic with Plato's eternal mind, describing the highest form of love as the intellectual love of unchanging ideas. Thomas Aquinas' notions of amicitia dei (divine-human friendship) and visio dei (beatific vision) provide an example of how our divine image can find its expression and fulfillment in the love of an eternal and unchanging God.

However, what is striking about the implications of our advancing knowledge of the human genome is by no means the unchanging or eternal. Instead it is the mammoth potential for altering the course of temporal and changing reality. We have to presuppose the idea of evolutionary change if we are even to conceive of the prospect of making an enduring improvement in individual human lives, let alone later generations in the human species. We are concerned here with the realm of time, not eternity. Furthermore, our focus is initially on the physical, not the mental, although our mental lives are necessarily influenced. Does this mean, then, that genetics research is simply irrelevant to religious conceptions of the imago dei? No. Its relevance forces us to examine another candidate for identifying the imago dei, namely, creativity. We think of God as the creator of the universe. We humans are creative as well. The big difference, of course, is that God creates from nothing, whereas we reshape or transform what already exists. Yet these two are not as radically different as it might at first seem. They can be combined. We can begin by conceiving of God's original act of bringing the universe into existence this way; God establishes a material world and then gives it a future, an open future in which new and different things might be made. And God's creative activity need not be limited to what He did at the point of origin. God's creative activity can be discerned as ongoing. God is at work everywhere all the time conserving, transforming, redeeming and renewing. In theological language, we can combine creatio ex nihilo or creation out of nothing with creatio continua or continuing creation.

We human beings are a product of this divine creativity, to be sure; but in turn we participate in the ongoing creative activity of opening up new futures. As homo faber we humans engage in constant shaping of our environment; and now perhaps we may even reshape our biological selves. Some speak of this as the human race beginning to exert a conscious influence on its own evolution. For instance, Philip Hefner, a theologian, at the Center for Religion and Science in Chicago, has begun to describe the human being as the "created co-creator," as totally dependent upon God yet a partner with God in opening up previously unknown possibilities for the future.

Posing these questions and suggesting that answers might be found in the notion of ongoing creativity could become significant as religious leaders wake up to what is going on in scientific research. The prospect of gaining the power to alter significantly the genetic make-up of future generations may initially evoke anxiety, fear of the unknown future and of our power to misshape it. This might provoke a defensive reaction, an immediate appeal to what is allegedly eternal and unchanging. It might provoke a defiant assertion that God's creation is fixed and unalterable. Genetic engineering will be dubbed a violation of something sacred, a transgression of something holy.

Ethical Questions: The Future of the Family

One of the most important ethical questions raised by the genome project concerns the future of the family. The question is, How will advancing knowledge of the human genome further erode the tie between marital union and procreation? There is little doubt that the project will increase the pressure to break the connection between sexual intercourse and procreation of offspring. In vitro fertilization and pre-embryo examination are already available for sex selection.

There are several issues tied together here that will have considerable impact on the future of the family. The first has to do with parent-child relationships. How essential is the tie between the parents' sexual life and their ability to care properly for their children? Will a genetically engineered child receive less care than one produced by passion? The widespread success of families with adopted children seems to testify that the tie between sex and procreation is far from fully decisive for insuring child welfare. Are there any theological reasons for conserving the tie between the unitive and pro-creative? One would hope that Roman Catholic moral theologians especially would not delay in giving attention to these issues.

The second involves discarding the nonimplanted proto-embryos. What is the moral status of destroying embryos on the basis of their undesirable genetic make-up? Are these fertilized eggs human? If so, is discarding them a form of abortion? Is "genecide" homicide? If this is a form of abortion, then the massive number of discarded proto-embryos will raise moral dilemmas to staggering proportions. We can expect that part of future abortion discussion will simply extend the debate we have heard since *Roe v. Wade*. What is new and in need of theological and ethical assessment is the discarding of unwanted pre-embryos on the grounds that they are genetically defective. This brings us to the next question: What does it mean to be genetically defective?

In a strictly scientific sense one might be able to construct a straightforward understanding of "defective." In cases involving the eye cancer retinoblastoma, for example, a slice of the DNA on chromosome 13 is deleted and moved over to chromosome 3. There is no doubt that the cancerous cell has a genetic defect.

But such a laboratory understanding is of minimal value in assessing the ethical and social consequences. Clinical criteria alone are not enough. The big question is this: On what basis do we decide who lives and who dies? How do we define "defective" in such a way that a defective life should be either prevented or terminated?

The problem [concerning employment and health issues] is that it is in the best interest of both insurance carriers and employers to evade the financial responsibility of paying for the health care of those at high risk. A new pattern of discrimination against persons with certain genetic codes may be in the offing. Genetic discrimination will reveal an incongruity that has long existed in the United States: the arbitrary tie between employment and health care. There is no logical connection. Yet in America one's access to health care is tied to one's job. If one has difficulty in the workplace, one's health care suffers accordingly. Currently there are an estimated thirty-five million Americans who have no health insurance and another fifteen million who are underinsured. The advent of genetic discrimination will continue the present incentive to make a profit from denying health care, and it will add new and more refined criteria so those at greatest risk are least likely to receive the health care they need.

The only solution may be a massive restructuring of economic incentives so that it will become profitable to provide health care to those who need it. One way to do this would be to break the tie between employment and medical services, something most employers would welcome. Some speculate that this might lead to a national rather than a strictly commercial program. Perhaps. One thing is obvious: unless pro-active political steps are taken, the medium range future will bring untold misery in health care and employment.

Source: Ted F. Peters and Robert J. Russell, "The Human Genome Project: What Questions Does It Raise for Theology and Ethics?" *Midwest Medical Ethics*, summer 1992, 12–17.

DOCUMENT 52: Philip Hefner, "The Evolution of the Created Co-Creator" (1989)

Some theologians are addressing the questions posed by genome research. Philip Hefner, a theologian at the Center for Religion and Science at the Lutheran School of Theology in Chicago, has formed answers of his own, viewing human beings as created co-creators.

* * *

[T]he evolving world process is actually a process of responding and adapting to God. God has created a process that is woven on the loom of adapting to its creator. . . .

... I believe a good case ought to be made for interpreting the entire physical and biological nonhuman prehistory as the instrumentality employed by God to introduce freedom into the cosmos and set up the introduction of the human being as the created co-creator. . . .

I recommend that we think of the human being as the *created co-creator*. Because we are *created*, we are reminded that we are dependent creatures. We depend for our very existence on our cosmic and biological prehistory; we depend on the grace of God. Yet, we are also *creators*, using our cultural freedom and power to alter the course of historical events and perhaps even evolutionary events. We participate with God in the ongoing creative process. In addition, the term "created co-creator" connotes the fact that we have a destiny. We have a future toward which we are being drawn by God's will. Only when we understand what this destiny is will we be able to measure and evaluate the direction we take in our creative activity.

Source: Philip Hefner, "The Evolution of the Created Co-Creator," in Ted Peters, ed., *Cosmos as Creation: Theology and Science in Consonance* (Nashville: Abingdon Press, 1989), 211–233.

DOCUMENT 53: Jeremy Rifkin, *Algeny: A New Word—A New World* (1984)

There are those, such as Jeremy Rifkin, director of the Foundation on Economic Trends in Washington, D.C., who oppose most current developments in biotechnology.

* * *

To end our long, self-imposed exile; to rejoin the community of life. This is the task before us. It will require that we renounce our drive for sovereignty over everything that lives; that we restore the rest of creation to a place of dignity and respect. The resacralization of nature stands before us as the great mission of the coming age.

Two futures beckon us. We can choose to engineer the life of the planet, creating a second nature in our image, or we can choose to participate with the rest of the living kingdom. Two futures, two choices. An engineering approach to the age of biology or an ecological approach. The battle between bioengineering and ecology is a battle of values. . . . If it is physical security, perpetuation at all costs, that we value most, the technological mastery over the becoming process is an appropriate choice. But the ultimate and final power to simulate life, to imitate nature, to fabricate the becoming process brings with it a price far greater than any humanity has ever had to contend with. By choosing the power of authorship, humanity gives up, once and for all, the most precious gift of all, companionship.

Can any of us, for that matter, entertain even for a moment the prospect of saying no to the age of biotechnology? If we cannot even entertain the question, then we already know the answer. Our future is secured. The cosmos wails.

Source: Jeremy Rifkin, *Algeny: A New Word—A New World* (New York: Viking Press, 1984), 252 and 255.

DOCUMENT 54: Ron Cole-Turner, *The New Genesis: Theology and the Genetic Revolution* (1993)

Reverence for life plays an important role in these philosophical and theological arguments. People question how we should view DNA and whether or not it should be treated as a sacred mystery. Theologian Ron Cole-Turner provided his answers, which reject the notion of treating DNA as something holy.

* * *

Once metal was thought to be sacred. . . . If for some philosophic or religious reason we believed that DNA is more sacred than any other complex chemical, then it would follow that we should be very cautious about acting directly upon it. But such a conviction of DNA's sanctity is not grounded in Western philosophic or religious traditions. Employing it now, in the context of genetic engineering, is arbitrary. To think of genetic material as the exclusive realm of divine grace and creativity is to reduce God to the level of restriction enzymes, viruses, and sexual reproduction. Treating DNA as matter—complicated, awe-inspiring, and elaborately coded, but matter nonetheless—is not in itself sacrilegious.

Source: Ron Cole-Turner, *The New Genesis: Theology and the Genetic Revolution* (Louisville, Ky.: The Westminster/John Knox Press, 1993), 45.

DOCUMENT 55: Statement from the National Council of Churches of Christ (1980)

Some religious groups have prepared statements that portray their support in favor of genetic research. One such statement was prepared by the National Council of Churches under the leadership of Roger L. Shinn.

* * *

The Christian churches understand themselves as communities dedicated to obeying the will of God through service to others. The churches have a particular concern for those who are hurt or whose faith has been shaken, as demonstrated by the long history of the churches in providing medical care. . . . Moreover, the churches have a mission to prevent suffering as well as to alleviate it.

Source: Human Life and the New Genetics, Report of a task force commissioned by the National Council of Churches of Christ. (New York: Office of Family Ministry and Human Sexuality), 47 (1980).

DOCUMENT 56: United Methodist Church Genetic Science Task Force, Draft Report to Annual and Central Conferences (1990)

Other religious groups, however, stress the need for caution in developing this new research and to think about its implications, particularly in relation to other needs.

* * *

Genetic technology makes possible the identification of the genetic basis of many common ailments. Scientists might soon develop tests for a genetic predisposition to many diseases that have both genetic *and preventable* environmental causes. In the short term, such a test might divert attention from finding ways to prevent the development of the disease. A propensity toward a particular disease might even become just one more "genetic disorder" for prospective parents to weigh in a parenting decision. . . . Also, should genetic screening be done in order to detect conditions for which there is no therapy?

Society has a responsibility to help shape the guidelines. For example, is it appropriate to develop genetic therapy for ailments better handled by improved diets? Or, is it appropriate at all to insert human genes into another animal even if that means leaner meat, and less heart disease? Is it appropriate to devote research to the development of leaner beef so that we can continue to eat meat with less adverse effects rather than devoting the research to the development of food products which can grow in areas of the world where starvation is rampant?

We affirm that knowledge of genetics is a resource over which we are to exercise stewardship responsibilities in accordance with God's reign over creation.

We caution against the concept of the technological imperative as an outgrowth of the prevalent principle in research that what *can* be done *should* be done.

We urge greater public funding and greater public control of genetic research.

Source: United Methodist Church Genetic Science Task Force, Draft Report to Annual and Central Conferences, December 1990, in *Christian Social Action*, January 1991, 19 and 21.

DOCUMENT 57: United Church of Christ Statement to the Seventeenth General Synod (1989)

Churches have also begun to address the issue of genetic discrimination, issuing statements concerning unfair discrimination in work, health care, insurance, and education.

* * *

Some have feared that genetic engineering confers too much power upon human beings, that we lack the wisdom to interfere with creation, or that we are "playing God." Some have argued that altering organisms is dangerous and the risks outweigh the benefits. Others recall the past misuse of human power in sinful ways and remind us that insidious prejudice still infiltrates some contemporary controversies over genetics. . . . These fears point to deep societal and religious concerns about human limits and our proper role in nature.

While general population screening may be used to protect individuals from danger, we reject screening as a basis for determining civil, economic, or reproductive rights.

Source: United Church of Christ Seventeenth General Synod, 1989.

DOCUMENT 58: Episcopal Church Statement to the Seventieth General Convention (1991)

There is no theological or ethical objection against the production and use of medicinal materials by means of genetic manipulation for the therapeutic or diagnostic purposes aimed at the prevention or alleviation of human suffering.

There is no theological or ethical objection against gene therapy, if proved to be effective without undue risk to the patient and if aimed at prevention or alleviation of serious suffering.

The benefits of this new technology should be equally available to all who need these for the prevention or alleviation of serious suffering, regardless of financial status.

The use of results of genetic screening of adults, newborns, and the unborn for the purpose of discrimination in employment and insurance is unacceptable.

Source: Seventieth General Convention of the Episcopal Church, 1991.

DOCUMENT 59: United Methodist Church Science Task Force, Report to the 1992 General Conference

We support the privacy of genetic information. Genetic data of individuals and their families shall be kept secret and held in strict confidence unless confidentiality is waived by the individual or his or her family, or unless the collection and use of genetic identification data are supported by an appropriate court order.

We support increased study of the social, moral, and ethical implications of The Human Genome Project. We support wide public access to genetic data that do not identify particular individuals.

We oppose the discriminatory or manipulative use of genetic information, such as the limitation, termination, or denial of insurance or employment.

Source: United Methodist Church Science Task Force, Report to the 1992 General Conference.

DOCUMENT 60: World Council of Churches, *Biotechnology* (1989)

The World Council of Churches does not directly acknowledge a belief that the embryo is a person from the time of conception, but it does affirm that the human embryo has some moral status.

* * *

Should we do important therapeutic research on an embryo which is not done for the benefit of the embryo itself and which even causes the destruction of the embryo? This problem raises enormously difficult questions about the status of the human embryo as a potential human being.

If we decide to perceive the human embryo not as a potential member of the human community but rather to be human biomass, then we are free to experiment on it without hesitation. If we concede that the potential of the human embryo to grow to one or even more individual persons requires our respect, then we are obligated only to experiment on it, if at all, for grave reasons. The difficulty lies in deciding what is or is not a sufficiently grave reason.

The basic theological question is if and how we recognize our neighbor in the living organism that evolves out of the merging human sperm and ovum. Following the principle of suspecting at least the potential for human development and therefore expanding protection rather than diminishing it, one may consider how the present standards of research on human beings could be adopted to embryo research. According to a declaration of the World Medical Association, dangerous and non-therapeutic research on human beings is allowed only for experimenting on one's self. This is impossible for embryos as well as it is for children.

Source: World Council of Churches, *Biotechnology: Its Challenges to the Churches and the World* (Geneva: World Council of Churches, 1989).

DOCUMENT 61: *Donum Vitae* (1987)

The Catholic Church supports some uses of prenatal diagnosis but is opposed to its use in conjunction with abortion. The genetic information in the gametes can be examined to determine genetic makeup, but after fertilization a different moral issue exists.

* * *

For prenatal diagnosis makes it possible to know the condition of the embryo and of the fetus when still in the mother's womb. It permits or makes it possible to anticipate earlier and more effectively, certain therapeutic, medical or surgical procedures.

Such diagnosis is permissible, with the consent of the parents, after they have been adequately informed, if the methods employed safeguard the life and integrity of the embryo and the mother, without subjecting them to disproportionate risks. But this diagnosis is gravely opposed to the moral law when it is done with the thought of possibly inducing an abortion depending on the results. A diagnosis which shows the existence of a malformation or a hereditary illness must not be the equivalent of a death sentence. Thus a woman would be committing a gravely illicit act if she were to request such a diagnosis with the deliberate intention of having an abortion should the results confirm the existence of a malformation or abnormality.

Source: *Donum Vitae*, Instruction on Respect for Human Life in Its Origin and on the Dignity of Procreation (Congregation for the Doctrine of the Faith, 1987), in Thomas A. Shannon and Lisa S. Cahill, *Religion and Artificial Reproduction* (New York: Crossroad, 1988), 150.

DOCUMENT 62: National Human Genome Research Institute, Policy on Availability and Patenting of Human Genomic DNA Sequence (1996)

As human genome research has progressed, policies have necessarily developed to address such issues as patenting genetic material. One such policy, developed by the National Center for Human Genome Research, addresses the availability and patenting of human genetic information obtained through grants funded by the government. The core of the policy is that the raw human genetic information should be released to the public as rapidly as possible and that such raw data are inappropriate for patenting. In addition, a monitoring period is established to determine the appropriateness of this policy.

* * *

Background

The Human Genome Project (HGP) is an international research effort, begun in 1990, which has the scientific goals of generating maps of the human genome and producing the complete sequence of the human DNA by the year 2005. The project was undertaken in the U.S. following the advice of several scientific committees that emphasized its importance in creating a resource that "will facilitate research in biochemistry, physiology and medicine," "have a major impact on health care and disease prevention" and provide "enormous scientific and technological advances . . . , having both basic and commercial applications." At NIH, the National Human Genome Research Institute (NHGRI) was founded to implement the HGP.

There are very strong scientific arguments that human genomic DNA sequence should be freely available and in the public domain:

—The human genomic DNA sequence is unique. Although there are many other types of information that contribute to the understanding of human biology, e.g., DNA sequence of model organism genomes, in the end, the only source of definitive information about the human is the human sequence.

—The human genomic DNA sequence is a vast resource. It contains a very large number of genes and an enormous amount of additional biological information. It is anticipated that the sequence resource will be the basis for many useful inventions and patentable products. It will take many researchers years to find and characterize all of the genes and other functional elements within the sequence and to use that information to develop products and other approaches that will improve the health of the American people.

—The human genome is a bounded resource. Once the genome has been sequenced, few or no opportunities will exist for discovery of new information that will not make reference to, or be dependent on, that first sequence. Thus, it is important to ensure maximum access of a large number of parties to the initial genomic DNA sequence as it is generated, to provide a broad opportunity for development of new products.

Policy

It is therefore NHGRI's intent that human genomic DNA sequence data, generated by the projects funded under RFA HG–95–005, should be released as rapidly as possible and placed in the public domain where it will be freely available. In order to implement this policy, NHGRI will require that grantees under RFA HG–95–005 adopt a policy of rapid release of data to public databases. This policy will be made a condition of the award.

In NHGRI's opinion, raw human genomic DNA sequence, in the absence of additional demonstrated biological information, lacks demonstrated specific utility and therefore is an inappropriate material for patent filing. NIH is concerned that patent applications on large blocks of primary human genomic DNA sequence could have a chilling effect on the development of future inventions of useful products. Companies are not likely to pursue projects where they believe it is unlikely that effective patent protection will be available. Patents on large blocks of primary sequence will make it difficult to protect the fruit of subsequent inventions resulting from real creative effort. However, according to the Bayh-Dole Act, the grantees have the right to elect to retain title to subject inventions and are free to choose to apply for patents should additional biological experiments reveal convincing evidence for utility. The grantees are reminded that the grantee institution is required to disclose each subject invention to the Federal Agency providing research funds within two months after the inventor discloses it in writing to grantee institution personnel responsible for patent matters. NHGRI will monitor grantee activity in this area to learn whether or not attempts are being made to patent large blocks of primary human genomic DNA sequence.

During this pilot period, NHGRI will be soliciting opinions and collecting evidence from the broad scientific and commercial sectors to allow an evaluation of whether the approach described above is sufficient to ensure that sequence generated by these grants is maximally useful to the research and commercial sectors. If not, NIH will consider a determination of exceptional circumstance to restrict or eliminate the right of parties, under future grants, to elect to retain title.

Source: National Human Genome Research Institute, Policy on Availability and Patenting of Human Genomic DNA Sequence Produced by NHGRI Pilot Projects (Funded under RFA HG–95–005), April 9, 1996.

DOCUMENT 63: Ted F. Peters, "Genome Project Forces New Look at Ethics, Law" (1993)

Several social consequences of the Human Genome Project have been and continue to be considered by those involved in the project. This document examines several of these issues. One is the possibility of genetic discrimination based on the knowledge that an individual has a predisposition for or a gene for a particular disease. Another area of importance is a potential for increased abortions because of an undesired genetic profile. Access to a genetic profile could also lead to efforts to modify that profile to enhance particular characteristics such as height or intelligence. The new genetics will enhance our capacities; we need to think through what this will mean for us and our society.

* * *

The new knowledge acquired through the Human Genome Initiative will require the human race to re-think its fundamental views on ethics, law, and society. That thinking already is taking place as research on the project unfolds.

Scientists realize that their work on the project places them in a world beyond science. When James D. Watson, who with Francis Crick discovered DNA's double-helix structure, asked the U.S. Department of Health and Human Services to appropriate funds for the Human Genome Initiative, he urged that 3 percent of the budget be allotted to study the research's ethical, legal, and social implications.

"We must work to ensure that society learns to use the information only in beneficial ways," he said, "and, if necessary, pass laws at both the federal and state levels to prevent invasions of privacy . . . and discrimination on genetic grounds." (1)

Watson was appointed first head of the Office of Human Genome Research at the National Institutes of Health (NIH). In a sign of things to come, he resigned last year amidst a dispute with NIH Director Bernadine Healy over the morality of patenting DNA sequences. (2) Healy and Watson have since been replaced by Francis Collins of the University of Michigan, who has been appointed director of the National Center for Human Genome Research. We no doubt can expect genetics and ethics to court one another for the next few years, leading perhaps to a marriage—to "genethics." (3)

The controversy that arose over recombinant DNA in the late 1970s and early 1980s precipitated an awareness among theologians and other church leaders that research in genetics may require ethical monitoring. Since then, a handful of theologians and ethicists have begun a dialogue with research scientists to sort through the issues.

GENETIC DISCRIMINATION

For three-quarters of all Americans, medical insurance is tied to employment. Twelve of Fortune's 500 top companies report that they genetically screen job applicants. In the past, screening was justified for public health reasons, but today it is increasingly used by employers to cut medical insurance premiums. Underwriters have denied or limited coverage to some gene-related conditions, such as sickle-cell anemia, atherosclerosis, Huntington's disease, Down's syndrome, and muscular dystrophy. (4) That list could increase. People with genetic dispositions to expensive diseases may become unemployable, uninsurable, and ultimately unable to obtain medical care. Geneticists estimate that each person carries five to seven lethal recessive genes, as well as a larger number of genes that make each of us susceptible to developing multifactorial diseases. In short, no one is disease free. This fact may be overshadowed by the breath-taking knowledge created by the Human Genome Initiative. Those screened in the early years are likely to suffer discrimination because of their apparent singularity. Only later, when we rediscover the relative equality of risk distribution, will the stigma fade. (5)

In the meantime, socially conscious ethicists may seek to ward off discrimination by invoking the principles of confidentiality and privacy. They will contend that genetic testing should be voluntary and the information contained in the genome be controlled by the patient. (6) The argument presumes that if information is controlled, the rights of the individual for employment, insurance, and medical care can be protected. Some legal grounds exist for this approach. Title VII of the 1964 Civil Rights Act restricts pre-employment questioning to work-related health conditions, and current state and federal legislation favors privacy. Nevertheless, the privacy defense is at best a stop-gap measure. Insurance companies will press for legislation more favorable to their interests, which is at odds with an individual's right to privacy. In addition, computer links make it difficult to prevent the movement of data from hospital to insurance carrier or to anyone else bent on finding out.

This game of hide-and-seek misses the main point, however. The key principle is this: The more we know and the more who know, the better health care planning can be. Thus, in the long run, what we want is information without discrimination.

The only way to reach this goal is to restructure the employment/insurance/health care relationship. The current structure makes it profitable for employers and insurance carriers to discriminate against individuals with certain genetic configurations; in short, it is in their financial interest to deny health care to those who need it the most.

A reformed health care system should make it more profitable to deliver, not withhold, health care. Such a system would require the nation to think of itself as a single community willing to care for all its own constituents.

[A multidenominational stand found among different Christian churches] provides a solid foundation from which to build an ethical proposal. There are hints that church ethicists will side with those who advocate privacy, and there are hints that they favor a national program that guarantees health care to everyone. What we do not yet see among religious leaders is an overall vision of the potential value (or lack of value) of widespread use of genome information for health care delivery.

ABORTION CONTROVERSY

The most divisive moral issue in the United States today is abortion. The advance of genetic knowledge and the development of more sophisticated reproductive technologies will only add nuance and subtlety to an already complicated debate.

Techniques have been developed to examine in vitro fertilized eggs as early as the fourth cell division to identify "defective" genes such as the chromosomal structure of Down's syndrome.

Prospective parents soon may be able to fertilize a dozen or so eggs in the laboratory, screen for the preferred genetic makeup, implant the desired zygote(s), and discard the rest. What, then, would be the status of the discarded pre-embryos? Might they be considered abortions? By what criteria do we define "defective"? Should prospective parents limit themselves to eliminating defective children, or should they go on to screen for desirable genetic traits, such as blue eyes or higher intelligence? Might this lead to a new form of eugenics, to selective breeding based upon personal preference and prevailing social values? What will become of human dignity in all this?

Challenges compound quickly. University of Texas law professor John A. Robertson says the central ethical issue in pre-implantation genetic screening is discarding unwanted embryos. Those who view the fertilized egg as the point at which human dignity and human rights begin contend that such embryos must be placed in the uterus and given an opportunity to grow. Those who hold this view undoubtedly would oppose the discarding of genetically defective embryos, just as they oppose abortion of fetuses with those defects. They are likely to support legislation that would prohibit discarding embryos.

In contrast, those favoring pre-implantation screening, with its accompanying discard of unwanted zygotes, contend that these embryos are too rudimentary in form or development to be owed respect because they are still undifferentiated cells not yet clearly individual. (7)

The legal precedent set by the landmark abortion ruling of *Roe vs. Wade* does not address the discarding of pre-implanted embryos. It legalized abortion as an extension of a woman's right to determine what happens to her body. This reasoning would not apply to pre-implanted embryos, however, because they are life forms outside a woman's body.

THERAPY VS. INTERVENTION

A third ethical issue, which appears in both secular and religious discussions, is the distinction between somatic therapy and germline enhancement. By "somatic therapy," we refer to the treatment of a disease in the body cells of a living individual by trying to repair an existing defect.

Body cells contain a full complement of 46 chromosomes. In contrast, germline cells in the reproductive process contain only 23 chromosomes, and these are passed on to the next generation. Intervention into the germline cells would influence heredity. The potential to significantly enhance the quality of life of future generations through germline intervention looms on the genetic horizon.

Ethical commentators largely agree that somatic therapy is morally desirable, and they look forward to the advances that the Human Genome Initiative will bring to this important work. Yet, they stop short of endorsing genetic selection and manipulation to enhance the quality of life for otherwise normal individuals or for the human race as a whole.

The Human Genome Initiative may locate genes that affect the brain's organization and structure. If this occurs, then careful engineering may enhance abstract thinking or other forms of physiological and mental improvement. (8) Such speculations are greeted with the greatest caution. Molecular hematologist W. French Anderson says: "In short, genetic enhancement risks violating human dignity by opening the possibility of discrimination." Religious ethicists agree: somatic therapy should be pursued, but enhancement through germline engineering raises serious questions about protecting human dignity.

Another reason for caution on germline enhancement, especially among Protestants, is the specter of eugenics, which connotes the ghastly racial policies of Nazism. This accounts for much of today's mistrust of genetic science in Germany and elsewhere. (9)

No one expects a resurrection of the Nazi nightmare, yet some critics fear a subtle form of eugenics slipping in the cultural back door. (29) The rising power to control the design of living tissue could foster an image of the "perfect child," and a new perception of perfection could begin to oppress all who fall short. Although not yet prominent in the literature, religious ethicists, speaking in March 1992 at a "Genetics, Religion and Ethics Conference" in Houston, Texas, saw the image of the "perfect child" to be a clear and present danger.

HUMAN DIGNITY

The list of ethical issues arising from the increasing ability of science to manipulate the basic building blocks of life itself is lengthy. It includes questions regarding the definition of a "defective" gene, equity of access to genetic services, gender justice, environmental impact, patenting new life forms, and developing biological weapons. Some farsighted religious leaders have entered into serious conversation with conscientious scientists so that cooperative thinking about our response and responsibility can be anticipated. (10)

Virtually all Roman Catholics and Protestants who take up the challenge of the new genetic knowledge agree on a handful of theological axioms. First, they affirm that God is the creator of the world and, further, that God's creative work is ongoing.

Second, the human race is created in God's image. In this context, the divine image in humanity is tied to creativity. God creates. So do we. With surprising frequency, humans are described as "co-creators" with God, making our contribution to the evolutionary process. To avoid the arrogance of thinking that humans are equal to the God who created them, some theologians have elected the phrase, "created co-creators," to describe human opportunity and responsibility. (11)

Third, these religious documents place a high value on human dignity. By "dignity," they mean what philosopher Immanuel Kant meant— namely, that we treat each human being as an end, not a means to some further end. The United Church of Canada eloquently voices the dominant view:

"In non-theological terms, it [dignity] means that every human being is a person of ultimate worth, to be treated always as an end and not as a means to someone else's ends. When we acknowledge and live by that principle, our relationship to all others changes." (12)

Pope John Paul II begins to appropriate the dignity principle in an elocution in which he condemns "in the most explicit and formal way experimental manipulations of the human embryo, since the human being, from conception to death, cannot be exploited for any purpose whatsoever." (13)

As church leaders respond responsibly to new developments in the Human Genome Initiative, we can confidently forecast one thing: the affirmation of dignity will become decisive for thinking through the ethical implications of genetic engineering.

Yet there is more. The theology of co-creation leads Ron Cole-Turner to a grand vision: "For the church, it is not enough to avoid the risks. Genetic engineering must contribute in a positive way to make the world more just and more ecologically sustainable, and it must contribute to the health and nutrition of all humanity." (14)

NOTES

1. James D. Watson. "The Human Genome Project: Past, Present, and Future," *Science* 248 (April 6, 1990), p. 46.

2. Healey instituted a plan to seek patents for NIH-funded discoveries to head off patents by private biotech companies. Watson objected, contending that the secrecy surrounding the patent process would restrict the free flow of scientific information necessary to facilitate research.

3. See Ann Lammers and Ted Peters, "Genetics: Implications of the Human Genome Project," *The Christian Century* 107 (October 3, 1990), pp. 868–872.

4. Mitchel L. Zoler, "Genetic Tests," *Medical World News* (January 1991), p. 14.

5. Alexander Morgan Capron, "Which Ills to Bear? Reevaluating the 'Threat' of Modern Genetics," *Emory Law Journal* 39 (Summer 1990), p. 690.

6. Invoking the principle of confidentiality to avoid job discrimination is the tack taken by the Joint Working Group on Ethical, Legal, and Social Issues (ELSI) of human genome research sponsored by the Department of Energy and the National Institutes of Health meeting at Los Alamos National Laboratory on April 29, 1991. Chaired by Nancy Wexler, the group issued a statement calling on the Equal Employment Opportunity Commission to revise its codes to grant greater privacy protections. See *Human Genome News* (September 1991), pp. 12–13.

7. John A. Robertson, "Procreative Liberty and Human Genetics," *Emory Law Journal* 39 (Summer 1990), p. 707.

8. Christopher Wills speculates on the physical enhancements and spiritual limits of Human Genome Initiative: "There is nothing in principle to stop us from eventually tracking down genes that influence intelligence, skills, mental health, even behavior. But there are no genes, on any of our 23 chromosomes, for the soul. . . . The mere fact that we cannot prove the existence of a soul, or assign it

a chromosomal location, does not invalidate that world view." See *Exons, Introns, and Talking Genes: The Science Behind the Human Genome Project* (NY: Harper Collins, Basic Books, 1991), pp. 314–315.

9. See Peter Meyer, "Biotechnology: History Shapes German Opinion," *Forum for Applied Research and Public Policy* 6 (Winter 1991), pp. 92–97.

10. See Troy Duster, *Backdoor to Eugenics* (NY: Routledge, 1990) and Jeremy Rifkin, Algeny (NY: Viking, 1983), esp. pp. 230–234.

11. In addition to ethical responsibility, church leaders also anticipate a broader pastoral responsibility. Ethicist Karen Lebacqz recommends that congregations respond to new developments in genetic therapy through (1) support for persons and families facing illness and death; (2) education of and by the clergy to help all church members interpret genetic illness and its accompanying hopes and tragedies; and (3) advocacy in public policy debate that will support legislation to aid disabled persons and others needing social protection. See Karen Lebacqz, ed., *Genetics, Ethics, and Parenthood* (NY: Pilgrim Press, 1983).

12. United Church of Christ, p. 2; World Council of Churches, p. 31; Methodist Church, p. 114; Church of the Brethren, p. 453; United Church of Canada, p. 14.

13. United Church of Canada, p. 14. See Philip Hefner, "The Evolution of the Created Co-Creator," in Ted Peters, ed., *Cosmos as Creation: Theology and Science in Consonance* (Nashville: Abingdon Press, 1989), pp. 211–234.

14. Pope John Paul II, "Biological Experimentation," October 23, 1982, *The Pope Speaks*, 1983.

15. Cole-Turner, "Genetics and the Church," p. 55.

29. See Troy Duster, *Backdoor to Eugenics* (NY: Routledge, 1990) and Jeremy Rifkin, *Algeny* (NY: Viking, 1983, esp. pp. 230–234.

Source: Ted F. Peters, "Genome Project Forces New Look at Ethics, Law," *Forum for Applied Research and Public Policy*, fall 1993, 5–13.

DOCUMENT 64: W. French Anderson, "Genetics and Human Malleability" (1990)

As noted in the previous document, somatic gene enhancement is becoming an issue of concern. One of the leaders in human gene therapy research, W. French Anderson, reflects on the need for great caution when it comes to engineering abstract thinking or other physiological and mental functions.

* * *

Somatic cell gene therapy for the treatment of severe disease is considered ethical because it can be supported by the fundamental moral principle of beneficence. It would relieve human suffering. Gene therapy would be, therefore, a moral good. Under what circumstances would human genetic engineering not be a moral good? In the broadest sense, when it detracts from, rather than contributes to, the dignity of man. . . . Somatic cell enhancement engineering would threaten important human values in two ways: It could be medically hazardous . . . [a]nd it would be morally precarious, in that it would require moral decisions our society is now not prepared to make, and it could lead to an increase in inequality and discriminatory practices.

Source: W. French Anderson, M.D., "Genetics and Human Malleability," *Hastings Center Report* 20 (January-February 1990): 23.

DOCUMENT 65: World Council of Churches, *Manipulating Life: Ethical Issues in Genetic Engineering* (1982)

The World Council of Churches has issued a similar statement on somatic cell enhancement.

* * *

[S]omatic cell therapy may provide a good; however, other issues are raised if it also brings about a change in germ line cells. The introduction of genes into the germline is a permanent alteration. Nonetheless, changes in genes that avoid the occurrence of disease are necessarily made illicit because these changes also alter the genetic inheritance of future generations. . . . There is no absolute distinction between eliminating defects and "improving heredity."

Source: *Manipulating Life: Ethical Issues in Genetic Engineering* (Geneva: World Council of Churches, 1982), 6–7.

DOCUMENT 66: Catholic Health Association, *Human Genetics* (1990)

The Catholic Health Association has a different view.

* * *

[G]erm-line intervention is potentially the only means of treating genetic diseases that do their damage early in embryonic development, for which somatic cell therapy would be ineffective. Although still a long way off, developments in molecular genetics suggest that this is a goal toward which biomedicine could reasonably devote its efforts.

Source: Catholic Health Association, *Human Genetics: Ethical Issues in Genetic Testing, Counseling, and Therapy* (St. Louis: Catholic Health Association of the United States, 1990), 19.

DOCUMENT 67: James D. Watson and Robert M. Cook-Deegan, "Origins of the Human Genome Project" (1991)

In 1991 James D. Watson and Robert M. Cook-Deegan published an article in the *FASEB Journal* that provided an overview of the steps that led to the development of the Human Genome Project. Their article concluded with a description of a unified and international genome project.

* * *

Robert Sinsheimer, then chancellor of the University of California at Santa Cruz, convened the first meeting to discuss the technical prospects for sequencing the human genome in June 1985 (1). This meeting did not have the intended result of establishing a research institute on campus but it did plant the idea of a large project in the minds of several prominent molecular biologists, most notably Walter Gilbert, Nobel laureate and professor of biology at Harvard who became the flag bearer for the genome movement for several months. He spoke about large-scale sequencing efforts at a Gordon Conference, a meeting on computers and molecular biology, and in other places in the late summer and fall of 1986.

EMERGENCE OF A CONCERTED GENOME PROJECT

Debate about the respective roles of the NIH and DOE dominated the policy concerns in 1988, with several congressional hearings devoted to that topic. A remarkable degree of joint planning began taking shape in 1989. Early in 1990, NIH and DOE submitted a 5-year joint research plan to Congress (2), an unprecedented accomplishment. The agencies established joint working groups on mapping, informatics (dealing with data bases and computational analysis), and social, ethical, and legal implications of genome research.

Many conferences and workshops are jointly sponsored, and the agencies are working toward a process to cross-refer certain kinds of grant applications, such as pilot sequencing projects and database management. NIH and DOE have agreed to notify one another about funding decisions regarding grant proposals reviewed by both agencies.

Much of the public debate during 1987 and 1988 centered on genome research programs at NIH and DOE. As programs at both agencies were successfully launched, the debate expanded. Other U.S. groups were also involved from early on: the Howard Hughes Medical Institute; the National Science Foundation (NSF) (without a genome program per se, but supporting instrumentation and genomic research on plants and non-human organisms); and the U.S. Department of Agriculture (focusing on plants of agricultural significance, and now considering extension to farm animals). HHMI has had a particularly strong interest in genetics for many years. HHMI has supported several human genetic databases as well as some of the best human genetics investigators in the United States. NSF has a long record of support for informatics, mathematical biology, and instrumentation, and coordinates a project on Arabidopsts thaliona.

The proliferation of research programs in different nations and agencies is a healthy sign of enthusiasm for genome research, laying the informational and technological foundations for a new approach to biology. Rapid growth and pluralistic sources of support, however, bring their own problems. Coordination of effort is preeminent among these. The central thrust of genome projects is to provide methods to accelerate biomedical research and to pool information about chromosomal structure. New methods must become rapidly dispersed and the information widely disseminated or there will be no point in special genome efforts.

FORMATION OF THE HUMAN GENOME ORGANIZATION AND UNESCO INVOLVEMENT

The Human Genome Organization (HUGO) was formed at the first Cold Spring Harbor meeting on genome sequencing and mapping on April 29, 1988. Victor McKusick provided the initial push, with strong support from Sydney Brenner, Leroy Hood, and James Watson. Brenner suggested the name, although he expressed a personal preference for THUG (The Human Genome Organization) as the acronym. McKusick was elected founding President, and HUGO began the arduous task of growing into an international scientific organization. In 1990, HUGO obtained commitments for several years' funding from the Wellcome Trust in the United Kingdom and the Howard Hughes Medical Institute in the United States. There were bright prospects for obtaining funds directly from various national scientific agencies for specified projects as the organization develops the infrastructure to support international workshops, fellowships, and other activities.

While HUGO was forming to mediate international scientific collaboration, the United Nations Educational, Scientific, and Cultural Organization (Unesco) evinced an interest in helping coordinate other aspects of international genome research. The center of Unesco's interest evolved toward ensuring continued input of third world nations in genome activities, and support for international conferences and scientist exchanges. The first major genome program under Unesco was a fellowship program, cosponsored by the third World Academy of Sciences, to provide limited funds for third world scientists to seek training in genome research centers throughout the world (V. Zharov, Unesco Science Office, personal communication, July 16, 1990).

It has taken 5 years for the genome project to be translated from an idea to the beginnings of an international scientific project. It is realistic to assume that it will continue to grow for the next several years. It is essential that the work continue, that the results be widely shared, and that governments of all developed nations contribute appropriately to the effort.

References

1. Sinsheimer, R. (1989) The Santa Cruz Workshop, May 1985. *Genomics* 5, 954–65.

2. U.S. Department of Health and Human Services and U.S. Department of Energy (1990) *Understanding our genetic inheritance: the first five years, FY 1991–1995.* DOE/ER–0452P. National Technical Information Service, Springfield, Virginia.

Source: James D. Watson and Robert M. Cook-Deegan, "Origins of the Human Genome Project," *FASEB Journal* 5 (January 1991): 8–11.

DOCUMENT 68: James D. Watson and Robert M. Cook-Deegan, "The Human Genome Project and International Health" (1990)

One of the major motives of mapping the human genome is its value for enhancing human health through the diagnosis and cure of various diseases. Researchers can be assisted in their crusade against disease and the development of means to understand the mechanisms of these diseases. The document that follows demonstrates how scientists are excited about the possibilities and positive outcomes of mapping the human genome. James Watson and Robert Cook-Deegan explore the international dimensions of genetic research and its potential benefits for all populations.

* * *

The human genome project is "public good" in the best sense. The purpose of the project is to construct common resources for the study of human genetics. The first fruits will be the development of several kinds of maps of the human genome, and those of other organisms, to permit rapid isolation of genes for further study about DNA structure and function. The major impact of the genome project on medicine will be a slow but steady conceptual evolution—a change in the way that we think about disease and normal physiology. A century ago, a revolution in medicine was in full stride following the discovery of infectious organisms and the dawn of bacteriology. Over the course of the century, the conceptual base of medicine has broadened from gross anatomy of organs to cellular biology to dissection of biochemical pathways. The next step is to study the most fundamental elements in biology —Mendel's hereditary factors now known as genes. . . . It is clear that in at least some cases, genetics will succeed where other approaches have failed.

The principal goal of the human genome project is to assist biomedical researchers in their assault on disease. The main benefit of genome research will be to provide tools to better understand the afflictions that exact an enormous toll of human suffering on every culture and in every geographic region.

Developed World

In developed countries, diseases that now kill and maim their victims are much less understood than the major infectious killing diseases of the turn of the century. Cancer, cardiovascular disease, stroke, Alzheimer's disease, arthritis, and autoimmune disorders involve complex interactions among environmental and genetic factors. Genetics will not explain everything, but having access to the genetic instruction set is essential to understanding any of these diseases in detail. Mapping and sequencing are first steps in that direction.

The genetics of cancer, for example, were largely unknown a decade ago. Much still remains a mystery, but the information derived from the cloning of genes and tracing their physiological impacts has been the most promising contribution to our understanding of cancer to date. The human genome project is merely a more global and systematic expansion of that approach.

The road to the new genetics is not yet paved. Genome-scale approaches to the study of human DNA are only beginning. The systematic approach to the study of monogenic diseases is the most immediate application, serving as the prototype for future research. But the scope of genetics is not restricted to disorders inherited in Mendelian fashion. Tools of genetics will prove far more generally applicable. In the past 2 years, techniques for cloning large DNA fragments of human chromosomes have been used to make detailed maps. On the X chromosome, for example, Schlessinger and colleagues have assembled over 1.5 million base pairs of contiguous DNA, cloned as yeast artificial chromosomes from the regions that surround both factor VIII and factor IX genes (1, 2) (which underlie two forms of hemophilia). This provides a resource for direct examination of DNA in that region. One group has proposed a pilot project to test the validity of sequencing the tip of the long arm of the X chromosome, which contains 30 disease-related genes. (1, 2) This illustrates the shift from the hunt for one gene at a time to the search for multiple genes by structural analysis of DNA.

Developing World

Complete maps and extensive sequencing data will have an impact on the diseases that plague the developing world, in addition to those diseases prevalent in developed nations. The most efficient way to diminish the toll of such diseases is most likely through improved nutrition, housing, sanitation and relief of poverty, but the effort against schistosomiasis malaria, Chagas' disease, and other disorders will not be complete without an understanding of the organisms that cause them. Understanding the genetics of an organism is fundamental to studying it scientifically, and the tools provided by genome research will make their contribution here. The same mapping approaches now used on yeast and nematodes could just as easily be applied to schistosomes, trypanosomes, and plasmodia, to study their life cycles at the most fundamental level, perhaps discovering new points of vulnerability and opportunities for better treatment.

Genome research will also generate methods and instruments applicable far beyond the mapping and sequencing of genomes. Methods based on direct analysis of DNA-hybridization, sequencing, polymerase chain reaction, and other techniques promise to revolutionize diagnosis. In the long run, DNA-based tests can be expected to supplant many of the measurements now used in clinical medicine. If such tests can be made simpler, faster, cheaper, and more accurate than existing tests, then more people can use them, including laboratories that are far from specialized as well as technologically sophisticated centers. If the genetic tools created by the genome project prove useful over the next 5 years, then special efforts to use them in the study of diseases prevalent in the developing world must receive serious attention. International support from government as well as private international funding organizations in North America, Europe, and Asia will doubtless be necessary for that task, as the populations most affected live in countries that lack the resources to finance such efforts.

Some populations in the Middle East, Asia, and Africa have been isolated by geography and limited technology, and represent rare and valuable resources to study human origins and patterns of population genetics. We must take steps, through international collaborations, to ensure that DNA is collected and stored for subsequent analysis, before these populations are lost for all time by the ingress of technology, the press of population growth, and other factors that break down the traditional barriers that have isolated these groups. This must be done with respect for the values and traditions of the populations to be studied.

Finally, if developing nations succeed in diminishing the impact of infectious disease, their health problems will come to more closely resemble those found in the developed world today. The genome project's benefits for diseases of the developed world are thus also relevant in the developing world, but the impact is obscured there by much larger and more urgent health issues.

International Genome Research Programs

The study of human genetics is now and always has been an international effort. The large-scale collaborations to study Duchenne's muscular dystrophy, Huntington's disease, Alzheimer's disease, cystic fibrosis, and almost any other disease that could be named involve collaborators from many nations and families spread throughout the world. The science is an international process, the sources of information respect no national boundaries, and the resulting information must be widely shared to be most useful.

Interest in genome research is clearly high throughout the world. The European Community began to discuss a genome research plan in late 1986.

Several national governments also have begun genome research programs. The US Department of Energy was first to start a dedicated genome research program in 1987. There are now two major government programs in the United States, at the National Center for Human Genome Research of the National Institutes of Health and at the US Department of Energy.

Italy mounted pilot genome projects early, soon after the US Department of Energy in 1987, and continues to support a $1.3 million per year genome program, soon to go up to $2 million per year. "The Medical Research Council in the United Kingdom announced plans for a 5 year, $11 million genome program that commenced in April 1989." The USSR announced a 25 million ruble genome program in June 1989, which has increased to 72 million rubles in 1990. There are also several genome-related research programs in Japan.

It is not yet clear how genome programs in various nations will work together. Ideally, coordination will be centered on scientific rather than political issues. A new international scientific organization, the Human Genome Organization (HUGO), was formed for this purpose at a Cold Spring Harbor, NY, meeting on genome research in April 1988. There are high hopes that HUGO can begin to operate as the "glue" that holds various national and private research efforts together. The Wellcome

Trust of the United Kingdom provided HUGO with its first large-scale funding of approximately £200,000 early in 1990, and other private philanthropies are likely to contribute as well. (3) This funding should enable HUGO to develop the capacity to convene periodic meetings, to hold scientific workshops focused on chromosomes or regions, to gather information needed to identify and resolve issues that require international action, and to track the progress of genome programs around the world.

Broad Social Implications of Genome Research

The power of the information to be gained from mapping and sequencing projects raises concerns about how it will be used. There is no avoiding the fact that arguments drawn in part from genetics have been politically misused in the past, most egregiously by the Nazis but also elsewhere in Europe and North America. (4) Indeed, the specter of coercive government eugenics programs persists even today in statutes still on the books in several nations, including some passed since the awful experiences of World War II. (5, 6) The only way to ensure that history does not repeat itself is for the scientific and medical communities to remain constantly vigilant for abuses of genetics.

The prospect of faster, cheaper, and more accurate genetic testing technologies has rekindled long-standing concern about how genetic patient data will be used, particularly for nonmedical purposes by third parties such as employers and insurers. Several recent books and government reports have dealt with concerns about genetic discrimination in applying for insurance, in workplace screening and testing, and in other contexts. (7) The public policy choices have not yet been made, and there should be vigorous debate about confidentiality of genetic information, the legitimacy of various uses, and access to important new genetic technologies. Scientists and clinicians should participate in this debate.

References

1. Cook-Deegan RM, Guyer M, Rossiter BJF, Nelson DL, Caskey CT. Large DNA cloning workshop. *Genomics*. In press.

2. Cook-Deegan RM, Rossiter, BJF, Engel L, Nelson DL, Caskey CT. Report of the X chromosome workshop. *Genomics*. In press.

3. Aldous P. Imperial's Wellcome support. *Nature*. 1990; 344: 5.

4. Kevles DJ. *In the Name of Eugenics*. Berkeley: University of California Press, 1985.

5. Wertz DC, Fletcher JC, eds. *Ethics and Human Genetics: A Cross-Cultural Perspective*. NY: Springer-Verlag, Inc., 1989. Chapter 2.

6. *Designer Genes: IQ, Ideology, and Biology*. Selangor, Malaysia: Instito Anaiisa Sosial, 1984.

7. Holtzman NA. *Proceed With Caution*. Baltimore, MD: Johns Hopkins University Press, 1989.

Source: James D. Watson and Robert M. Cook-Deegan, "The Human Genome Project and International Health," *Journal of the American Medical Association* 263 (June 27, 1990): 3322–25.

Part V

Issues in Research

As the pace of discoveries in genetics increases at an almost exponential rate, so too does the need for careful research into the application of these discoveries to humans. Even though the discoveries appear to be beneficial and offer the promise of cure or prevention of many diseases, care must be taken to show that there are no harmful side effects or other risks posed by the intervention. Thus there is the need to consider carefully the ethical issues posed by the application of new genetic knowledge and interventions to humans.

DOCUMENT 69: Thomas H. Murray, "Ethical Issues in Human Genome Research" (1991)

As research in the field of genetics has increased, ethical concerns have also increased. Thomas Murray provides us with a concise definition of bioethics and what he feels are its effects on human genome research.

* * *

Scientific research into human genetics has been a continuing source of intriguing, and at times formidable, ethical issues. . . .

Field of bioethics

A precise definition of ethics may be difficult to find, but there is considerable agreement on what is meant by bioethics. Bioethics is the study of ethical issues in medicine, health care, and the life sciences. As a formal field of study bioethics is approximately 20 years old, although a few pioneering scholars were writing in the 1950s on issues that were later incorporated into bioethics. From its earliest stirrings, bioethics has paid considerable attention to human genetics, wrestling with such issues as prenatal genetic testing and abortion, genetic manipulation, and eugenics. In the past 5–10 years, the field of bioethics has grown rapidly, expanding the range of questions scholars ask, and even experiencing controversy over method (1).

From the standpoint of bioethics, research on the human genome presents no completely novel ethical questions, at least for now. That is partly because of the nature of new ethical questions, which typically are variants of ethical questions that scholars and others have wrestled with before. This embeddedness of questions in experience with analogous questions means that we do not have to invent every response totally anew, but rather can draw on the history of scholarly analysis that has come before. The acceleration of knowledge about human genetics promised by genome research assures that the ethical questions presented will be plentiful and significant. They may be grouped into three categories: 1) the possibility of greatly increased genetic information about individuals and populations; 2) the manipulation of human genotypes and phenotypes; and 3) challenges to our understanding of ourselves, individually and collectively.

Although we are utterly unprepared to deal with issues of mandatory screening, confidentiality, privacy, and discrimination, we will likely tell ourselves that we have already dealt with them well . . . (p. 20) (2).

REFERENCES

1. Jonsen, A. R., and Toulmin, S. (1988) *The Abuse of Casuistry. A History of Moral Reasoning.* University of California, Berkeley

2. Annas, G. (1989) Who's afraid of the human genome? *Hastings Center Report* 19 (July–August): 19–20

Source: Thomas H. Murray, "Ethical Issues in Human Genome Research," *FASEB Journal* 5 (January 1991): 55–60.

DOCUMENT 70: American Society of Human Genetics, "Statement on Cystic Fibrosis Screening" (1989)

Ethical difficulties are already beginning to arise with the concept of presymptomatic testing for disease. People cannot be tested to find out if they have a genetic predisposition for particular diseases. Problems arise when testing involves those who may not wish to know whether they are predisposed to a particular disease. Other problems involve accuracy rates and whether or not proper counseling and follow-up help are being provided. Such problems have prompted professional bodies to issue statements, as the American Society of Human Genetics has done concerning testing for cystic fibrosis.

* * *

The current test detects only 70% of carriers, and there is little experience in the delivery of such complex information to large populations.

First, carrier testing should be offered to couples in which either partner has a close relative affected with CF [cystic fibrosis]. Second, one or a few federal, foundation, or privately supported pilot programs should be conducted as soon as possible in order to gather more data regarding laboratory, educational, and counseling aspects of screening. Third, there is an immediate need for centralized quality control of laboratories conducting these tests. Fourth, it will be appropriate to begin large-scale population screening in the foreseeable future, once the test detects a larger proportion of CF carriers and more information is available regarding the issues surrounding the screening process. Until that time, it is considered premature to undertake population screening.

Finally, while it is recognized that testing of highly motivated individuals in the general population may occur, it is the position of The American Society of Human Genetics that routine CF carrier testing of pregnant women and other individuals is NOT yet the standard of care in medical practice.

Source: American Society of Human Genetics, "Statement on Cystic Fibrosis Screening," *American Journal of Human Genetics* 46 (November 1989): 393.

DOCUMENT 71: American Society of Human Genetics, "Statement on Genetic Testing for Breast and Ovarian Cancer Predisposition" (1994)

A similar statement was made by the American Society of Human Genetics concerning testing for breast and ovarian cancer predisposition.

* * *

Once direct and reliable testing for BRCA1 mutations is available, it may be offered to members of specific types of families with strong breast-ovarian cancer histories. While the cancer risks associated with different BRCA1 mutations are being determined, testing should initially be offered and performed on an investigational basis by appropriately trained health care professionals who have a therapeutic relationship with the patient and are fully aware of the genetic, clinical, and psychological implications of testing, as well as of the limitations of existing test procedures. Until then, it is recommended that linkage analysis be offered to selected high-risk families, if it will provide for more refined counseling than is currently available from family history alone.

Further research is needed to determine optimal monitoring and preventive strategies (surgical or chemo-prophylactic), to assure their efficacy.

It is premature to offer population screening, until the risks associated with specific BRCA1 mutations are determined and the best strategies for monitoring and prevention are accurately assessed.

Source: American Society of Human Genetics, "Statement on Genetic Testing for Breast and Ovarian Cancer Predisposition," *American Journal of Genetics* 55 (1994): i–iv.

DOCUMENT 72: J.B.S. Haldane, *Heredity and Politics* (1938)

Fears concerning the use and misuse of genetic research and testing, especially concerning discrimination in the workplace, have evolved from historical misuse of such information. Early work conducted by the geneticist J.B.S. Haldane, in which he concluded that people's responses to toxins in the workplace were in some part genetically determined, could and possibly did lead to this kind of discrimination.

* * *

But while I am sure that our standards of industrial hygiene are shamefully low, it is important to realize that there is a side to this question which has so far been completely ignored. The majority of potters do not die of bronchitis. It is quite possible that if we really understood the causation of this disease, we should find that only a fraction of potters are of a constitution which renders them liable to it. If so we could eliminate potters' bronchitis by rejecting entrants into the pottery industry who are congenitally disposed to it. We are already making the attempt to exclude accident-prone workers from certain trades. This principle could perhaps be carried a good deal further.

There are two sides to most of these questions involving unfavorable environments. Not only could the environment be improved, but susceptible individuals could be excluded. Thus my father's work on divers' paralysis and caisson disease among compressed air workers led not only to the drawing up of tables for the safe ascent of divers, but to the elimination of unduly fat men, who are particularly susceptible. It must be added that at present it is generally practicable to improve the environment, while we are very rarely able to discover what types of men are susceptible. Nevertheless, in a society which was based on a knowledge of biology it would be realized that large innate differences exist, and men would not be given tasks to which they were congenitally unsuited. We must no more forget heredity when we are trying to improve environment than we must forget environment when trying to improve heredity.

Source: J.B.S. Haldane, *Heredity and Politics* (New York: W. W. Norton, 1938), 192–193.

DOCUMENT 73: U.S. Congress, Office of Technology Assessment, *The Role of Genetic Testing in the Prevention of Occupational Disease* (1983)

Compulsory workplace genetic testing poses ethical problems because individuals should be the ones who decide whether they want to be tested and whether they want to work at a particular job if it is determined that they are at greater risk for developing particular diseases. This autonomy is taken away when workplaces decide whether to employ predisposed individuals. Some people also fear that the workplace may just decide not to hire those at greater risk rather than spending the time and money to reduce hazardous conditions. The U.S. Congress's Office of Technology Assessment studied workplace genetic screening in the early 1980s and concluded that the present state of genetic testing and research did not justify the screening of employees.

* * *

Although this technology [genetic testing] is still in its infancy, it has the potential to play a role in the prevention of occupational diseases. It is technologically and economically impossible to lower the level of exposure to hazardous agents to zero. However, if individuals or groups who were predisposed to specific types of occupational illness could be identified, other preventive measures could be specifically directed at those persons. This is the promise of genetic testing. At the same time, however, the technology has potential drawbacks and problems. For example, the ability of the techniques to identify people who are predisposed to occupational illness has not been demonstrated. In addition, some people are concerned that its use could result in workers being unfairly excluded from jobs or in attention being directed away from efforts to reduce work place hazards.

While it may be too soon to be able to answer many of the questions raised by genetic testing, it is not too soon for society to begin to consider them. The technology is developing, and some major companies have used it to a limited degree. Many more companies have expressed an interest in using it in the future. Moreover, genetic testing is one of a number of technologies that purport to identify people, both in and out of the workplace, who face an increased risk for disease. Policy decisions is made on issues raised by genetic testing are likely to be relevant to the issues raised by those other technologies.

Source: U.S. Congress, Office of Technology Assessment, *The Role of Genetic Testing in the Prevention of Occupational Disease* (Washington, D.C.: U.S. Government Printing Office, 1983).

DOCUMENT 74: The Office of Technology Assessment's Report on Genetic Monitoring and Screening in the Workplace (1991)

One option examined was the prohibition of genetic testing in the workplace.

* * *

POLICY ISSUES AND OPTIONS FOR CONGRESSIONAL ACTION

While technologies associated with genetic monitoring and screening in the workplace have continued to advance, OTA [Office of Technology Assessment] found no significant change in the use of these technologies since 1983. Thus, several of the policy issues and options for congressional action offered in the 1983 OTA report are still valid and remain unchanged.

Two central issues related to genetic monitoring and screening in the workplace were identified during the course of this assessment. They are:

—the appropriate role of the Federal Government in the regulation, oversight, or promotion of genetic tests (both monitoring and screening); and

—the adequacy of federally sponsored research on the relationships between genes and the environment.

Associated with each policy issue are several options for congressional action, ranging in each case from taking no specific steps to making major changes. Some of the options involve direct legislative action. Others involve the executive branch but with congressional oversight or direction. The order in which the options are presented does not imply their priority. Moreover, the options are not generally mutually exclusive; adopting one does not necessarily disqualify others that pertain to the same or other issues, although changes in one area could have repercussions in others. A careful combination of options might produce the most desirable effects.

ISSUE: Is there a role for the Federal Government in genetic monitoring in the workplace?

Option 1: Take no action.

Congress could take no action to prohibit, regulate, or promote the use of genetic monitoring in the workplace. This would allow employers, employees, and their representative trade groups and unions to regulate its use through negotiation, arbitration, and litigation.

Thus far, executive agencies involved in workplace health and safety have not regulated against the use of genetic monitoring in workplace settings. OSHA [Office of Safety and Health Administration] has regulated some employer practices that could affect the use of genetic monitoring, such as medical records access by the employee. Congress could take no action if it determines that present Federal regulation is adequate in this area. Under this scenario, constraints on the use of genetic monitoring would develop through court filings in suits between parties or by negotiations between companies and unions.

Option 2: Prohibit genetic monitoring in the workplace.

To prevent all possibilities for discrimination and breach of confidentiality, Congress could prohibit genetic monitoring in the workplace. In light of the many discrete changes needed in the OSH Act, NLRA [National Labor Relations Administration], and Rehabilitation Act to achieve this degree of protection through regulation, Congress could decide to prohibit all genetic monitoring until further research into the methods is conducted. Such a prohibition could shift the focus of the issue to levels of exposure in the workplace.

Prohibiting genetic monitoring, however, will delay the accumulation of data needed to make the judgment whether certain genetic monitoring tests are useful. By slowing the development of these data, prohibition might threaten efforts to identify workplace hazards, whether they are to be randomized through cleanup or worker selection. In addition, some workers who might have avoided dangerous exposures had they known of a susceptibility will sicken unnecessarily. Finally, this option clearly eliminates the possibility for mandatory worker protection under those discrete circumstances where overall work site hazard reduction is not technologically or economically feasible.

Option 3: Promote genetic monitoring in the workplace.

Congress could decide that genetic monitoring in the workplace should be promoted because of its potential to improve the work environment and worker health conditions. This could be done by providing additional funding to those Federal agencies currently performing research into genetic monitoring methods, as well as basic research on the cause of occupational disease, in general, and the relationships between environmental exposures and health effects arising from genetic mutation. Such projects could identify useful occupational genetic monitoring tests and develop protocols for their use. However, many questions about the use of genetic monitoring remain unanswered. Because the interpretation of genetic monitoring is only considered to be reliable at the population level, rather than the individual level, the current usefulness of genetic monitoring in the workplace is questionable.

ISSUE: Is there a role for the Federal Government in genetic screening in the workplace?

Option 1: Take no action.

Congress could choose to take no action in the area of genetic screening in the workplace. As with genetic monitoring, any constraints on the use of genetic screening would develop through court rulings in lawsuits between employers and employees, or by negotiations between companies and unions. In support of this option is the viewpoint that congressional action is not currently warranted at this time. Use of genetic screening in the workplace has not changed greatly since the 1983 OTA report.

However, there have been several newly recognized susceptibilities to occupational illness since that time. In addition, advances have been made in the area of molecular techniques for genetic screening for both occupationally and nonoccupationally related disease.

If Congress takes no action in this area, those identified as susceptible to occupational illness through genetic screening could be seen as unfit for work. In addition, those identified as being susceptible to a nonoccupationally related disease could be seen as a health insurance burden. Without proper restrictions the use of genetic screening to detect either type of disease risk could make job discrimination a possibility.

Option 2: Prohibit genetic screening in the workplace.

The principal reason for prohibiting genetic screening in the workplace would be the concern over its potential misuse. Such potential for misuse probably would be greater for genetic screening than genetic monitoring because the former is targeted toward identifying individuals at increased risk while the latter focuses on groups at increased risk. The existing legal framework may offer protection in some circumstances, but many questions have not been resolved.

A drawback to this option is that by prohibiting both types of genetic screening in the workplace, employers could not utilize screening for occupationally related disease. This type of screening offers some protection of the worker.

Option 3: Promote genetic screening in the workplace.

Congress could stimulate the research and development of genetic screening tools by providing funds for research into occupationally related diseases, nonoccupationally related disease, or both. Useful screening tests for occupationally related disease could be developed which could have direct benefits for the individual worker and indirect benefits for the employer. In addition, more money could be provided in the area of research on occupationally related traits. NIOSH [National Institute for Occupational Safety and Health] could be authorized to do research in this area, and to certify procedures for medical technologies that are of sufficient value to be used in an occupational setting. If such research were promoted, however, employers might be prone to use those screening tests to screen out susceptible employees in lieu of cleaning up the workplace.

If research in the area of nonoccupationally related disease was promoted beyond current Federal research levels, employers might be prone to using such screening tests to discriminate against employees or job applicants who might increase the company health care costs.

ISSUE: Should the Federal Government regulate genetic monitoring or screening in the workplace?

If Congress determines that the current regulatory framework addressing genetic monitoring and screening is adequate, it could take no action and let the current regulatory framework stand. However, if Congress determines that the current regulatory framework is inadequate, it could pursue several avenues. A framework established by several major pieces of legislation exists on which to build: OSH Act; NLRA; Title VII of the Civil Rights Act of 1964; Rehabilitation Act of 1973; and ADA [Americans with Disabilities Act]. The following options are discussed according to whether they apply to genetic monitoring, genetic screening, or both.

Genetic Monitoring and Screening

Option 1: Congress could amend Section 6(b)(7) of the OSH Act which states that OSHA standards shall prescribe the type and frequency of medical examinations or other tests to be made available, to specify whether genetic monitoring and screening tests are to be included.

To either prohibit or promote genetic monitoring and screening, Congress could amend this section's coverage with respect to genetic monitoring and screening. To contain abuses, Congress could include language directing OSHA to prescribe or recommend genetic monitoring or screening only when less intrusive medical tests will not provide information of substantially the same value. Thus, for example, tests for sickle cell anemia would not be permitted unless other tests of lung function and blood oxygenation were incapable of giving an employer the information needed to decide whether a particular worker could safely manage a particular task.

A principal drawback of this option is that performing genetic monitoring and screening tests on employees could be financially prohibitive for some employers. In addition, mandating genetic monitoring or screening tests could be burdensome for both the employer and the employee. Such an action could require the employer to hire new medical staff to perform them. Furthermore, the employee might not wish to undergo genetic monitoring or screening.

Option 2: Congress could amend the OSH Act to guarantee the confidentiality of genetic monitoring and screening results.

Congress could amend the OSH Act to specifically guarantee that genetic monitoring and screening results not be disseminated, except in nonidentifying, statistical forms for research purposes, to any third-party without specific authorization from the worker. Further, employers could receive only the conclusion of the occupational physician, i.e., whether the worker is fit for the job in question, without receiving details or results of the genetic monitoring or screening tests. The worker, on the other hand, would receive both the test results and the conclusions drawn from them by the examining physician. Several State statutes provide a model for such legislation.

Advantages of this option are the ability to shield workers from misuse of genetic information by immediate and potential employers, and the maintenance of adequate authority to provide statistical information needed for ongoing improvement of health and safety practices. As with the option just mentioned concerning recordkeeping, however, this amendment would logically be appropriate to all medical records, and not merely those concerning genetic monitoring and screening tests. Thus, evaluating this option requires a larger consideration of whether the OSH Act should guarantee the confidentiality of all medical testing in this fashion. If this option is adopted, consideration would also need to be given to remedies for breach of confidentiality and an examination of the role of the occupational physician employed by the company.

Option 3: Require full disclosure to employees and job applicants of the nature and purpose of all medical procedures performed on them.

Current law does not require employers to disclose the nature and purpose of medical procedures conducted on employees or job applicants, or how the results are to be used. Although employees are given access to their medical records, they may not be able to interpret the data within the records, or challenge incorrect information. A congressionally mandated requirement that employers provide detailed information of what procedures were performed and why they were performed might serve as a deterrent to abuses. This would also protect the employees' autonomy by allowing them to be part of a decisionmaking process that affects their health and economic interests. If the test were genetic in nature, the assistance of a genetic counselor would be important to fully explain the procedure and the meaning of a positive result.

On the other hand, by requiring full disclosure, Congress would place requirements on employers that might be perceived as burdensome and expensive. Additionally, arguments might be made that such a requirement would intrude on the judgment of the occupational health physician.

Genetic Monitoring

Option 4: Congress could direct OSHA to clarify that genetic changes shall be included under the definition of occupational illness.

OSHA's definition of occupational illness now includes "abnormal condition," but does not specifically cover genetic changes. Taking this action could ensure that data on worker exposures and subsequent genetic changes would be recorded in worksites where employers are using genetic monitoring. This would help with ongoing efforts to assess the effects of potentially hazardous substances, as well as offer the opportunity to more closely monitor the health of a particular worker.

Yet including genetic changes in the definition of occupational illness would implicitly equate all genetic changes with "illness." Many changes are likely to be without immediate symptomatic effect. Therefore, gathering and distributing this information might be unduly alarming, particularly to the workers in question. Also, all genetic change is not definitely a result of the workplace. Changes can be induced by personal habits and lifestyle decisions (e.g., smoking, diet) as well. Equating genetic changes with illness may encourage employers to view such employees as somehow disabled or unfit for work, making job discrimination a distinct possibility.

Genetic Screening

Option 5: Congress could amend section 504 of the Rehabilitation Act to prohibit discrimination in hiring against otherwise qualified applicants because their genetic screening results reveal a proclivity toward certain diseases in the future.

Amending section 504 in this manner would address several potential concerns. First, it tackles the problem of discrimination against job applicants, a topic left largely untouched by the OSH Act and NLRA protections. Second, it addresses what is perhaps the most likely area of abuse for the use of genetic screening. Third, it focuses on one of the possible uses of genetic screening, i.e., identification of applicants who are qualified but likely in the future to suffer from a disease that will require full use of sick leave or even early retirement. Finally, amending section 504 also permits Congress to address the use of genetic screening in the workplace to detect nonoccupationally related illnesses.

By focusing on this section, rather than with section 503, Congress could avoid the problem of directing employers to include those with genetic variants that do not otherwise qualify them as "handicapped" under their affirmative action programs.

The disadvantage, however, is the uncertainty associated with section 504's requirement that employers provide a reasonable accommodation for handicapped workers. It may be clear what accommodation is necessary to make a job accessible to one who is deaf or blind, and in turn to make a judgment whether that accommodation is reasonable to require of an employer. It may be more difficult, however, to judge what is necessary for someone with a currently asymptomatic genetic illness or susceptibility.

> *Option 6: Congress could direct the National Labor Relations Board to make preemployment genetic screening a mandatory subject of bargaining, in order to increase the possibilities for workers to protect themselves against what they and their representatives perceive as abuses of genetic screening.*

Fitness-for-duty physicals and medical tests are already regarded as mandatory subjects of bargaining between unions and employers when applied to current workers. Thus, extending the concept to preemployment physicals and genetic screening would not require markedly different concerns to be placed on the bargaining table, and would provide some protection for job applicants.

A disadvantage to this option, however, is that unions do not represent the majority of American workers, so this action would not protect all affected persons. Additionally, in light of interest in "two-tiered" systems of compensation, it is possible that unions and employers may trade protections for current workers from potentially discriminatory genetic screening tests for the privilege of screening job applicants more stringently.

Source: U.S. Congress, Office of Technology Assessment, *Medical Monitoring and Screening in the Workplace: Results of a Survey—Background Paper*, OTA-BP-BA-67 (Washington, D.C.: U.S. Government Printing Office, 1991).

DOCUMENT 75: Council for Responsible Genetics, Position Paper on Genetic Discrimination (1997)

The Council for Responsible Genetics also has a position on the use of genetic information.

* * *

Genetic testing is not only a medical procedure. It is also a way of creating social categories. As a basic principle, we believe that people should be evaluated based on their individual merits and abilities, and not based on stereotypes and predictions about their future performance or health status. In most cases, genetic testing can only reveal information about probabilities, not absolute certainties. We believe that individuals should not be judged based on stereotypes and assumptions about what people in their class or status are like.

Insurance or employment practices which employ these stereotypes in underwriting inadvertently reinforce them in other arenas as well. There is a strong public policy precedent for avoiding the negative social consequences of such a practice.

Skin color, like other genetic traits, is mediated by genes. These lie entirely outside the individual's control. Whereas individuals can exercise choices about whether to smoke, how much exercise they get, and how much fat is in their diets, they cannot change the contents of their genes. To make employment or insurance decisions on the basis of genetic characteristics determined at the moment of conception is to discard cherished beliefs in justice and equality.

Source: Council for Responsible Genetics, Position Paper on Genetic Discrimination, August 18, 1997.

DOCUMENT 76: Nelson A. Wivel and LeRoy Walters, "Germ-Line Gene Modification and Disease Prevention: Some Medical and Ethical Perspectives" (1993)

Germ-line gene modification for the prevention and treatment of diseases raised many questions among medical professionals and ethicists. One such highly debated form of intervention involves the transfer of properly functioning genes into reproductive cells. The ethical issues that arise as a result of this type of research are outlined in the following document.

* * *

There has been considerable discussion about the merits and risks of germ-line gene modification in humans (1). Previous publications make it clear that this is a topic that readily provokes debate (2, 3). [We will consider] some of the complex ethical quandaries that necessarily attend decisions about deliberate alteration of the human germ line.

There are numerous ethical arguments for and against germ-line gene modification, and the following discussion is meant to be representative of the types of issues raised by various observers. Although it would be unrealistic to expect a consensus with regard to the most compelling arguments of either persuasion, it seems apparent that most arguments have been influenced by the underlying culture, in this, a pluralistic Western democratic society, with a strong interest in individual rights.

There are several arguments in favor of developing the capabilities for human germ-line genetic intervention (Table 1).

1) The health professions have a moral obligation to use the best avail-

able methods in preventing or treating genetic disease, and certain types of genetic disorders may require germ-line alterations. If the current strides in the study of molecular genetics continue and sometime in the future it is appropriate to consider germ-line gene modification, it should be done in the context of extending a therapeutic continuum; chemicals have already been given to activate dormant genes, such as the attempted use of 5-azacytidine to increase fetal hemoglobin production in patients with sickle cell anemia, and lung transplantation has been used to treat cystic fibrosis (3). When it is determined that germ-line intervention is the best application of molecular medicine for a given disease, and if it is acceptably safe and efficacious, then it will be in the best interests of patients to have the health care system offer this technology. To rule out this option in advance and in principle would mean breaking with a long-standing tradition of medicine to either treat or prevent all types of diseases (4).

2) The principle of respect for parental autonomy should permit parents to use this technology to increase the likelihood of having a healthy child. With the burgeoning growth of in vitro fertilization and the increasing ability to make compensatory maneuvers for various types of infertility, there has been a strong declaration of parental autonomy, ultimately directed toward optimizing the chances for the birth of a healthy child. Notwithstanding certain court decisions regarding abortion or the request to terminate life support, it would be ethically problematic for legislators or judges to interfere with procreative liberty when parents are acting on the basis of their deeply held moral convictions and are attempting to prevent disease in their offspring through germ-line modification (5).

3) Germ-line gene modification is more efficient than the repeated use of somatic cell gene therapy over successive generations. At least in the case of highly prevalent genetic disorders, disease prevention through germ-line modification may be the most efficient approach to reducing the incidence of disease. For example, although five protocols have been approved by the NIH Recombinant DNA Advisory Committee for somatic cell gene therapy of cystic fibrosis, none of these interventions would affect more than the patients selected for the particular studies. Because cystic fibrosis is the most common genetic deficiency disease among U.S. Caucasians, occurring in approximately 1 in 2500 births, one could easily make the case that it would be more efficient and cost-effective to use germ-line gene modification to eliminate the problem both for the patient and for future generations (6).

4) The prevailing ethic of science and medicine operates on the assumption that knowledge has intrinsic value and should be pursued in the vast majority of cases. The acquisition of knowledge is of fundamental importance to science and medicine. The mere fact that advances in

gene targeting or preimplantation diagnosis could lead in the future to proposals for germ-line modification in humans should not deter researchers from pursuing these lines of inquiries. Similarly, if germ-line modification becomes a reliable technique for preventing disease in laboratory animal models, the fear of possible misuse of the technique in humans should not interfere with the conduct of well-controlled clinical trials with human subjects. It would be both inhumane and tragic if special interest groups or those who set public policy attempted to block potentially promising lines of scientific inquiry on the basis of political viewpoints or speculative fears. Reasonable public policies will attempt to prevent the misuse of new technologies while also having the goal of promoting the development of novel approaches to the prevention and cure of serious disease.

Table 1. Some ethical arguments for and against germ-line gene modification.

Arguments in favor

Moral obligation of health professions to use best available treatment methods.

Parental autonomy and access to available technologies for purposes of having a healthy child.

Germ-line gene modification more efficient and cost-effective than somatic cell gene therapy.

Freedom of scientific inquiry and intrinsic value of knowledge.

Arguments against

Expensive intervention with limited applicability.

Availability of alternative strategies for preventing genetic diseases. Unavoidable risks, irreversible mistakes.

Inevitable pressures to use germ-line gene modification for enhancement.

Any balanced perspective on germ-line intervention must acknowledge that numerous arguments against the use of this technology in humans have also been advanced. Some of the more important counterarguments (Table 1) are the following:

1. Germ-line gene modification is an expensive intervention that would affect relatively few patients. The incidence of classical Mendelian genetic diseases is quite low; the sum total of known genetic diseases affects approximately 2% of all live births. To invoke the need for a very expensive technology for the attempted remediation of extremely uncommon situations is a difficult matter at a time when there is public discussion of rationing medical resources. Although the expense of pre-embryo diagnosis is considerable, there is no reason to assume that germ-line gene modification would be less expensive, particularly be-

cause the latter process could require large numbers of oocytes and embryos to assure the safety and efficacy of the germ-line approach.

2. Alternative strategies exist for avoiding genetic disease. The need for germ-line gene modification may be avoided by improved strategies of preimplantation and prenatal genetic diagnosis. Embryo freezing has been successfully demonstrated in animal models and in humans (7). Blastomeres can be removed from the embryo and retain their viability (8). With the aid of polymerase chain reaction analysis, the DNA from one or two cells can be amplified and subjected to diagnostic tests for multiple genetic and chromosomal disorders (9). Such techniques have made possible the genetic testing of embryos before implantation. Preimplantation diagnosis represents an advance over the methods of amniocentesis and chorionic villus biopsy, which require a pregnancy to be in place and for which the only currently available intervention is selective abortion. Selection processes such as these can achieve many of the goals of germ-line intervention with procedures that may have a lower order of risk. However, whether to transfer or discard embryos that are heterozygous carriers of a genetic trait remains an important issue for the selection strategy.

3. The risks of the technique will never be eliminated, and mistakes would be irreversible. Germ-line gene modification will always be associated with the risk of predictable genetic side effects, and for this reason it never should be approved in humans. Whatever the medical review and approval, they are not likely to be fail-safe because it is not possible to guarantee safety and reproducibility in biological systems. Further, there is the present potential for the delayed appearance of unpredicted side effects that can be passed onto future generations; for example, subtle adverse effects on them could appear many years after genetic intervention, and such effects might not be detected in animal models that were used to develop the preclinical data. In sum the risks are much greater than those cited with somatic cell gene therapy where the side effects are most likely confined to one patient.

4. Germ-line gene modification for serious disease will inevitably lead to the next step, genetic enhancement (10). Germ-line gene modification is a dangerous step onto a "slippery slope." Although initial emphasis of this type of genetic alteration may be on the prevention of disease, it seems likely that there would follow a gradual shift to include efforts at enhancement (11). It is true that there are both clinical researchers and bioethicists who have asserted that therapy can be differentiated from enhancement in a definitive way (12). Indeed, on the surface it would appear reasonably straightforward to set up a rather defined dichotomy between the use of germ-line intervention for prevention of and use for enhancement. However, maintaining this dichotomy could prove a difficult task. There already exist precedents for treating conditions that

would not meet any consensus definition of disease. For example, the treatment by means of recombinant human growth (HGH) of dwarfism secondary to growth hormone deficiency has not been a provocative step, but a recent decision to administer recombinant HGH to children of short stature, who have no evident HGH deficiency, has been highly criticized. The criticism has centered around the thesis that short stature, per se, is not a disease (13) and that this intervention therefore is an enhancement rather than a medically indicated treatment.

REFERENCES AND NOTES

1. H. J. Muller, *Perspect. Biol. Med.* 3, 1 (1959); G. Wolstenholme, Ed., *Man and His Future* (Churchill, London, 1963); Council of Europe, Parliamentary Assembly, 33rd Ordinary Session, "Recommendation 934 (1982) on Genetic Engineering" (Strasbourg); President's Commission for the Study of Ethical Problems in Medicine and Biomedical and Behavioral Research, *Splicing Life* (Government Printing Office, Washington, DC, 1982); J. C. Fletcher, *Va. Law Rev.* 69, 515 (1983); U.S. Office of Technology Assessment, *Human Gene Therapy: Background Paper* (Government Printing Office, Washington, DC, 1984); G. Fowler, E. T. Juengst, B. K. Zimmerman, *Theor. Med.* 10, 151 (1989); E. T. Juengst, *J. Med. Philos.* 16, 587 (1991); M. Lappe, ibid., p. 621; K. Nolan, ibid., p. 613; Canada Law Reform Commission (B. M. Knoppers), *Human Dignity and Genetic Heritage: Study Paper* (Law Reform Commission, Ottawa, 1992); S. Elias and G. J. Annas, in *Gene Mapping: Using Law and Ethics as Guides*, G. J. Annas and S. Elias, Eds. (Oxford Univ. Press, New York, 1992), pp. 142–154; R. Munson and L. H. Davis, *Kenn. Inst. Ethics J.* 2, 137 (1992).

2. J. D. Howell, *Clin. Ethics* 2, 274 (1991); R. Moseley, *J. Med. Philos.* 16, 641 (1991); A. L. Caplan, in *Gene Mapping: Using Law and Ethics as Guides*, G. J. Annas and S. Elias, Eds. (Oxford Univ. Press, New York, 1992), pp. 128–141; B. D. Davis, *Hum. Gene Ther.* 3, 361 (1992); J. C. Fletcher and W. F. Anderson, *Law Med. Health Care* 20, 26 (1992); J. V. Neel, *Hum. Gene Ther.* 4, 127 (1993).

3. H. I. Miller, *Hum. Gene Ther.* 1, 3 (1990).

4. B. K. Zimmerman, *J. Med. Phil.* 16, 593 (1991).

5. R. M. Cook-Deegan. ibid., p. 163.

6. L. Walters, *Nature* 320, 225 (1986).

7. A. Trounson and L. Mohr, ibid., 305, 707 (1983).

8. R. S. Prather and N. L. First, *J. Exp. Zool.* 237, 347 (1986).

9. M. W. Bradbury, L. M. Isola, J. W. Gordon, *Proc. Natl. Acad. Sci. U.S.A.* 86, 3709 (1989).

10. D. J. Weatherall, *Nature* 331, 13 (1988), ibid., 349, 275 (1991).

11. E. M. Berger and B. M. Gert, *J. Med. Philos.* 16, 667 (1991); A. Mauron and J.-M. Thevoz, ibid., p. 648.

12. W. F. Anderson and J. C. Fletcher, *N. Engl. J. Med.* 303, 1293 (1980); W. F. Anderson, *J. Med. Philos.* 14, 681 (1989); *Hastings Cent. Rep.* 20 (no. 1), 21 (1990); L. Walters, *Hum. Gene Ther.* 2, 115 (1991).

13. A. Toufexis, *Time* 142, 49 (1993).

Source: Nelson A. Wivel and LeRoy Walters, "Germ-Line Gene Modification and Disease Prevention: Some Medical and Ethical Perspectives," *Science* 262 (October 22, 1993): 533–538.

DOCUMENT 77: American Society of Human Genetics, Ad Hoc Statement on Insurance Issues in Genetic Testing (1995)

Places of employment aren't alone in considering the benefits of genetic research in screening employees. Those individuals who know they are at higher risk for developing specific diseases are more likely to buy more insurance in anticipation of needed care. This will lead to competition among insurance companies, which will encourage places of employment to screen for predisposition. Insurance companies, then, are also examining questions of genetic testing and discrimination in applications for insurance, as this statement from the Ad Hoc Committee on Insurance Issues in Genetic Testing of the American Society of Human Genetics shows.

* * *

Distinguishing and classifying individuals at different risks is at the heart of commercial insurance as it is practiced today, particularly when individual and small group policies are involved. Insurers do not believe that genetic conditions—or genetic tests that predict illness, death, or disability—should be excluded from this traditional practice. Differentiation of applicants on the basis of health risks is legal and should be distinguished from discrimination which is illegal if based on race, gender, or sexual orientation.

The extent of such legal discrimination at present is not clear, nor can its future course be predicted with any clarity. There have been reports of denial of coverage on the basis of genetic information, though some have questioned the validity of many of these claims.

At present it appears that insurance companies do not require genetic tests, particularly molecular tests, in underwriting. This is partly because of both the rarity of most genetic disorders and the high cost of such tests, as well as unfamiliarity with their validity and usefulness in the insurance setting. However, "multiplexing" of tests could result in marked reductions in cost, so that genetic testing could become as commonplace as multiphasic chemistry tests or multiple tests on a urine sample. Although insurers may not now require genetic testing, they nevertheless do make decisions based on genetic information, including family history or prior diagnostic tests performed in the course of delivery of medical care to the applicant and his or her family.

Source: American Society of Human Genetics, "Genetic Testing and Insurance," *American Journal of Human Genetics* 56 (1995): 327–331.

DOCUMENT 78: President's Commission for the Study of Ethical Problems in Medicine and Biomedical and Behavioral Research, *Splicing Life: The Social and Ethical Issues of Genetic Engineering with Human Beings* (1982)

One widely debated use of genetic research is gene therapy, in which diseases are prevented or cured by correcting or replacing defective genes. Both the U.S. Presidential Commission and the U.S. Congress used reports to address this concern in 1980.

* * *

Genetic engineering techniques are already demonstrating their great potential value for human well-being. The aid that these new developments may provide in the relief of human suffering is an ethical reason for encouraging them.

—Although the initial benefits to human health involve pharmaceutical application of the techniques, direct diagnostic and therapeutic uses are being tested and are already in use. Those called upon to review such research with human subjects, such as local Institutional Review Boards, should be assured of access to expert advice on any special risks or uncertainties presented by particular types of genetic engineering.

—Use of the new techniques in genetic screening will magnify the ethical considerations already seen in that field because they will allow a larger number of diseases to be detected before clinical symptoms are manifest and because the ability to identify a much wider range of genetic traits and conditions will greatly enlarge the demand for, and even the objectives of, prenatal diagnosis.

Many human uses of genetic engineering resemble accepted forms of diagnosis and treatment employing other techniques. The novelty of gene splicing ought not to erect any automatic impediment to its uses but rather should provide thoughtful analysis.

—Especially close scrutiny is appropriate for any procedures that would create inheritable genetic changes; such interventions differ from prior medical interventions that have not altered the genes passed on to the patients' offspring.

—Interventions aimed at enhancing "normal" people, as opposed to remedying recognized genetic defects, are also problematic, especially since distinguishing "medical treatment" from "nonmedical enhancement" is a very subjective matter; the difficulty of drawing a line suggests the danger of drifting toward attempts to "perfect" human beings once the door of "enhancement" is opened.

Source: President's Commission for the Study of Ethical Problems in Medicine and Biomedical and Behavioral Research, *Splicing Life: The Social and Ethical Issues of Genetic Engineering with Human Beings* (Washington, D.C.: U.S. Government Printing Office, 1982).

DOCUMENT 79: President's Commission, *Screening and Counseling for Genetic Conditions* (1983)

Another report by the President's Commission also spoke directly to ethical issues in genetic screening.

* * *

Genetic information should not be given to unrelated third parties, such as insurers or employers, without the explicit and informed consent of the person screened or a surrogate for the person.

Mandatory genetic screening programs are only justified when voluntary testing proves inadequate to prevent serious harm to the defenseless, such as children, that could be avoided were screening performed. The goals of "a healthy gene pool" or a reduction in health costs cannot justify compulsory genetic screening.

Source: President's Commission for the Study of Ethical Problems in Medicine and Biomedical and Behavioral Research, *Screening and Counseling for Genetic Conditions* (Washington, D.C., U.S. Government Printing Office, 1983).

DOCUMENT 80: National Institutes of Health Recombinant DNA Advisory Committee's Subcommittee Statement: "Gene Therapy for Human Patients" (1989)

The National Institutes of Health will not fund gene therapy research that does not pass many levels of review, including a special subcommittee of the Recombinant DNA Advisory Board. This subcommittee has a specific document outlining the requirements that must be satisfied before research can be approved.

* * *

What disease do you intend to treat with gene therapy?

Why do you consider this disease to be an appropriate candidate for treatment with these new methods?

What laboratory studies have been done, with cells and live animals, that make researchers hopeful that gene therapy will help patients rather than harming them?

What are the probable benefits and harms of the proposed treatment, both to the patient and to others?

If there are several patients who need gene therapy but only one of them can be treated initially, how will selection be made in a way that treats all patients fairly?

How will patients—or in the case of young children, the parents of patients—be properly informed about the possible benefits and risks of gene therapy?

What steps will be taken to protect the privacy of the patient and the patient's family, while at the same time informing the public about the results of gene therapy?

While the RAC (Recombinant DNA Advisory Committee) and its Subcommittee believe that gene therapy for non-reproductive, or somatic, cells holds promise for patients suffering from certain genetic and other diseases, they will seek to ensure that patients are not subjected to unreasonable risk or harm, excessive discomfort, or unwanted invasion of privacy and that they will receive special care, monitoring, and consideration. The public will be informed about every step that is taken with this new technique.

Source: Human Gene Therapy Subcommittee of the NIH Recombinant DNA Advisory Committee, "Gene Therapy for Human Patients: Information for the General Public" (Bethesda, MD: National Institutes of Health, 1989).

DOCUMENT 81: Michael S. Langan, "Prohibit Unethical 'Enhancement' Gene Therapy" (1997)

Recent changes in policy no longer require privately funded research to undergo scientific or ethical review by the Recombinant DNA Advisory Committee and only scientific review by the FDA. The National Organization for Rare Disorders opposes this position.

* * *

While we can rest assured that gene therapy "enhancement" experiments will not move forward immediately, or at least until adequate vectors are developed, we must move swiftly to set parameters around this issue before the human gene pool is altered irrevocably. In addition, the barriers erected to mark the point beyond which scientists should not go, must affect federally funded research as well as the private sector. NORD's observation of the field thus far indicates that gene therapy will continue to move in the direction of the most profitable human conditions because there is even far more money to be made in curing baldness and wrinkles than there ever will be in cancer or HIV/AIDS.

The danger is the *absence* of regulation over moral and ethical issues. Previously, the RAC was able to raise these issues as it reviewed individual protocols, but it is no longer empowered to deny approval of experiments even if they are ethically repugnant. If "enhancement" protocols move forward—to make a person taller, or thinner, or change the color of their eyes or skin—there will be no way to stop them as long as the FDA determines that they are safe and effective. Moreover, the RAC's scope of responsibility has always been confined to publicly funded research, therefore the private sector can freely pursue enhancement therapies without any public oversight, as long as no federal funds are involved.

The National Organization for Rare Disorders urges Dr. Varmus to return protocol approval authority to the NIH Recombinant DNA Advisory Committee.

Source: "Prohibit Unethical 'Enhancement' Gene Therapy," Statement of Michael S. Langan, Vice President of Public Policy, National Organization of Rare Disorders, Inc., to the NIH Gene Therapy Policy Conference (Bethesda, MD): September 11, 1997.

DOCUMENT 82: Donald M. Bruce, "Moral and Ethical Issues in Gene Therapy" (1996)

The Church of Scotland, which sponsors an ongoing Science, Religion, and Technology Project, has spoken against germ-line or somatic-cell therapy.

* * *

The precautionary principle enshrined in the current UK legal prohibition of germline and early embryo gene therapy is certainly appropriate at our current state of genetic knowledge. . . . It is quite possible that there never could obtain such knowledge as would change this situation, or that the cost would make it impractical. What is more difficult is the ethical question of whether we would ever have the right to make genetic decisions on behalf of future generations. Some would argue on principle that this violates individual rights unacceptably. Still others might argue that we have a duty to future generations to seek to eradicate genetic diseases wherever we reasonably can, but, as discussed above, this seems unrealistic in practice. It would seem hard and harsh to rule out absolutely any possibility of a specific future gene therapy that meant some extremely serious condition was not passed on to one's children, but the seriousness of contemplating such an irrevocable step is daunting. The "lesser of two evils" approach exemplified above, for a very few extreme clearcut cases, and perhaps involving only dominant genetic disorders, would seem tenable for a Christian in theory. In practice, it would be extremely hard to know at what point we could be confident that we were justified in proceeding in their best interests, acting as they would have wished us to do, were they present to say so.

Source: Donald M. Bruce, "Moral and Ethical Issues in Gene Therapy" (Church of Scotland Society, Religion, and Technology Project, 1996).

DOCUMENT 83: Jerome Kagan, "The Realistic View of Biology and Behavior" (1994)

Genetic research is used not only to examine the causes and possible cures for physical diseases but also to examine our psychological makeup. Jerome Kagan, a psychologist at Harvard University, explores this use of genetic research and provides his feelings on the matter.

* * *

The campaign to suppress discussion of biology grew strong during the opening decades of this century. Politically liberal scholars, joined by journalists of like mind, wished to mute the arguments of conservatives who argued for halting immigration from Eastern Europe on the ground that the immigrants had genetic flaws. The liberals were helped by Ivan Pavlov's discovery of conditioning, in the early 1900's, in a St. Petersburg laboratory. If a dog could be taught to salivate at a sound, surely a child could be taught anything, was the message John Watson, America's first behaviorist, brought to American parents after World War I. That bold claim was congruent with Sigmund Freud's creative hypothesis that family experiences in the early years could create or prevent a future neurosis.

By the late 1920's, the broad acceptance of inherited temperamental traits, which had lasted for two millennia, had been banished, its demise speeded by our society's need to believe in the power of social experience.

As the discipline of psychology—born in Europe during the last quarter of the 19th century—became recognized at American colleges and universities, many faculty members began to emphasize the influence of social experience on behavior. This approach became easier to defend after Hitler proposed the repugnant philosophy that Aryans were superior to other people.

After World War II, social science in America also became more positivistic, demanding objective evidence for all theoretical statements. Neuroscience was still young in the late 1940's and was unable to supply evidence that could explain, for example, how a particular physiological profile might be the foundation of an anxious or an angry mood.

By the 1970's, however, the historical context had again changed. Hundreds of studies of the way families influence growing children had not produced the powerful generalizations that had been anticipated a half century earlier. Equally important, engineers and scientists had invented ingenious ways to study the brain. Suddenly it became possible to speculate about how a particular neurochemistry could produce excessive anxiety, sadness, or anger. Scientists who put forward such explanations were not treated as intellectual terrorists for suggesting, for example, that a woman with panic attacks might have inherited a neurochemistry that rendered her especially vulnerable to a sudden, inexplicable sharp rise in heart rate, a feeling of suffocation, and a surge of fear.

But I believe that some psychiatrists and neuroscientists are moving too quickly toward a biological determinism that is as extreme as the earlier loyalty of some psychologists to an environmental explanation of behavior. Fortunately, a majority of scientists recognize that no human psychological profile is a product of genes alone.

At the moment, two psychological categories, which can be observed clearly in children, appear to be heavily influenced by biology. Between 15 per cent and 20 per cent of a large group of healthy infants we studied, who were born into secure homes, were found to have inherited a tendency to be unusually aroused and distressed by new, unexpected events. When they were observed in our laboratory at four months of age, they thrashed their limbs and cried when they saw colorful, moving mobiles or heard tape recordings of human speech. About two-thirds of these easily aroused infants, whom we call "high reactive," became extremely shy, fearful, subdued toddlers. Based on other data, we estimate that about one-half of this group of toddlers will become quiet, introverted adolescents. Not all of the high-reactive infants will become introverted, however, because their life experiences lead them to develop differently.

A second, larger group of infants—about 35 per cent of the children we studied—are the opposite of the high-reactive, shy children. These infants are relaxed and rarely cry when they experience new events. Two-thirds of this group become sociable, relatively fearless young children. Stressful events, however, can produce a fearful or shy manner in such children, even though they began life with a minimally excitable temperament.

Support for a biological contribution to the development of these two types of children comes from the fact that the two groups differ in many aspects of their physiological functioning, as well as in their body build. The fearful children show larger increases in heart rate and blood pressure when they are challenged—signs of a more reactive sympathetic nervous system. They have a higher prevalence of the allergies that produce hay fever and hives (and, surprisingly, possess narrower faces). Studies of identical and fraternal twins support the belief that each of these temperamental types is influenced, in part, by heredity.

. . . [W]e who study the relationship of biology to behavior must make clear that psychological phenomena, like a fearful or a fearless style of behavior, cannot be reduced completely to a person's biology. The child's life history influences the adult's psychological profile. Because the course of that life history is unknown when children are very young, we cannot at that time select the very small proportion of the 5 per cent to 10 per cent of children whose temperaments dispose them to be fearless, impulsive, and aggressive and will go on to develop an asocial or criminal personality. It would be unethical for example to tell parents that their 3-year-old son was at serious risk for delinquent behavior.

Similar arguments can be made about predicting which children will develop panic attacks, depression, or schizophrenia. A small group of children are at risk for each of these disorders because of the physiology they inherit, but we are unable at the present time to say which of the children will eventually develop a particular disorder—because we do not know what vicissitudes life will hold for them.

Perhaps future discoveries will supply the information that will make such predictions accurate enough to warrant benevolent intervention early in the child's life. We will have to wait and see whether that promise can be fulfilled.

A more subtle implication of the research on temperament involves people's willingness to take responsibility for their own actions. I trust that most Americans still believe in the notion of free will—that we can decide what action we will take or not take—and that each of us has a moral obligation to be civil and responsible, even when we wake up feeling blue, angry, or anxious. Our culture still insists that we should pull on our socks and act responsibly, even if that postulate comes at some emotional price.

The danger in the new romance with biology is that many people will begin to award temperament too strong a voice, deciding, for example, to be permissive and accepting of friends who lose their tempers too easily. Each of us does inherit a temperamental bias to one or more characteristics, but we also inherit the human capacity for restraint. Most of the time, humans are able to control the behavior that their temperament presses upon them, if they choose to do so. The new research on temperament and biology should not be used to excuse asocial behavior. Rather, the purpose of the inquiry is to help us understand the bases for the extraordinary variation in human motivation, moods, and social behavior.

We would do well to remember that although the poet-philosopher Lucretius believed in temperamental variation, he was also convinced that the "lingering traces of inborn temperament that cannot be eliminated by philosophy are so slight that there is nothing to prevent men from leading a life worthy of the gods."

Source: Jerome Kagan, "The Realistic View of Biology and Behavior," *The Chronicle of Higher Education*, October 5, 1994, 64.

DOCUMENT 84: Henry T. Greely, "Genes, Patents, and Indigenous Peoples: Biomedical Research and Indigenous Peoples' Rights" (1996)

Research has also led to the genetic analysis of indigenous peoples, another practice that raises ethical concerns. Henry T. Greely, professor of law at Stanford University, presents arguments in favor of this practice, explaining the benefits of this type of research as well as delineating the ethical concerns involved.

* * *

Scientists are now able to detect different forms of proteins and to "read" the stretches of DNA that order them made. These so-called "coding" regions of DNA constitute the bulk of the 50,000 to 100,000 "genes" in humans. The coding regions make up about five percent of the human genome: about 150,000,000 "letters" of the three billion or so in each person's genetic code. By learning the sequence of a gene, scientists can begin to understand better when, where, and how the protein made by the gene works—and what happens when the proper protein does not appear at the right time and place. Ultimately, scientists hope to be able to use this improved understanding of protein function to prevent or to cure many illnesses.

What does this have to do with indigenous peoples? It does not involve the existence of "special" genetic variations ("alleles") in indigenous populations. Genetically, humans are a very homogeneous species. Of the three billion "letters" in our individual genetic codes, two people from anywhere in the world will differ, on average, at one place in a thousand. In the "coding" sections of the genome, those differences fall to one in ten thousand. The genetic variations that do exist are generally found in all groups, although at higher or lower frequencies in some populations. There are no "genes" for being Chinese, or Navajo, or Zulu, or Irish, although genetic variations that affect skin color, hair color, nose shape, and the few other superficial characteristics associated with traditional views of "race" will be found in different frequencies in different populations. The notion that an ethnic group, indigenous or not, has "superior" genes is often quite tempting to members of that group. It has no scientific basis and, quite literally, is racist.

So why are some scientists interested in the genetic analysis of indigenous peoples? Many are interested not in functioning genes, but in what the patterns of genetic variation can reveal about a community's history and society. This interest extends far beyond indigenous populations, however defined. Almost all the detailed work that has been done in human genetics has been done with samples taken from people of European descent, not because of any conscious or subconscious racism but because most of the research has been done in Europe and North America. The result, however, has been research on "the" human genome that has largely excluded genetic variation in general and almost entirely ignored the variation found in about 85 per cent of humanity. The proposed Human Genome Diversity Project (HGDP) is one, resolutely non-commercial, effort to collect and make available, in a scientifically and ethically coordinated manner, genetic information about a representative sample of the world's entire human population: European and non-European, indigenous and non-indigenous.

Limited Commercial Prospects

There is no "gold rush" for indigenous DNA. The HGDP is important not because indigenous populations are being eagerly sought out by commercially-funded genetic researchers, but because these populations, along with most non-European populations, have largely been ignored. There are three limited circumstances, however, where genetic information from isolated populations could be medically, and commercially, important (1) common disease-linked genetic variations, (2) uncommon disease-linked genetic variations, and (3) possible "health-linked" genetic variations.

In order to learn the sequence of a gene associated with a particular disease, the scientists have to find the gene. Generally, that has involved finding families of people who have had a disease that seems to be linked to genes—the disease passes through families in particular ways and without any obvious, purely environmental cause. When families that have the genetic disease are found, the DNA of family members with the disease can be compared with the DNA of family members who do not have the disease. Stretches of DNA that are shared by sick family members but not by healthy ones are a good place to look for genes that are connected to the disease. The process is not easy: it can take hundreds of families, years of work, and scores of blind alleys, but its end results may be the location of the relevant gene, a copy of the gene, and, ultimately, the sequence of the gene itself.

Therefore, in order to find the gene, scientists first must find families in which the genetically-linked disease is common. Those families can sometimes be found in distinctive populations, some of which have been particularly useful in such research, for reasons indicated in the following discussion of research in the genetics of mental illness in one discrete group, the Amish.

The Old Order Amish community has large, extended families, and information on their ancestry is well known and available—genealogical records can trace their ancestry back to only 30 or so European progenitors. In addition to information on ancestry, medical and hospital records with demographic details are also available. The group is located within a small geographical location, and even individuals who leave the community tend to remain in the immediate area, making them accessible for genetic studies. The group is both genetically and culturally homogeneous, thereby reducing the variables that can complicate genetic studies. In addition, cultural taboos essentially eliminate alcohol or drug abuse, which is important for the study of affective disorders since the symptoms of these disorders can be masked by alcohol or drug use. Furthermore, individuals within this community have very well-defined roles, and close interactions between its members mean that behavior that differs from the norm is readily detectable. Finally, this community, as a whole, is very concerned about health issues and particularly about mental illness. They have therefore been extremely cooperative with researchers interested in understanding the basis for such disorders. (Risch and Botstein 1996)

For these reasons and others (such as physical proximity to researchers), the Old Order Amish have been studied often for information about

some genetically-linked diseases. Other populations that have been particularly important in genetic research include Mormon families from the
western United States and Finns.

Sometimes, indigenous populations will contain families that suffer
from a high rate of a genetically-linked disease. Those families can then
be studied in an effort to find the gene. Thus, researchers have long been
studying families in the Pima tribe (also sometimes called, with the Pa-
pago, the Tohono O'odham) in Arizona. For reasons that are thought to
combine genetic predisposition and current diet, the Pima suffer from a
very high rate of adult onset diabetes, a disease that maims and kills
millions of people of all ethnic backgrounds. There are numerous other
research efforts, ongoing or planned, looking at adult onset diabetes in
narrowly-defined populations, including the Finns, the Old Order
Amish, the "Gullah" Islanders in the American Southeast, and some
West Africans. Research on such common genetically-linked diseases has
the potential to provide information about the diseases that will benefit
people in every ethnic group, and, as a result, has the potential for large
commercial returns. At the same time, the studied populations are not
thought to have any unique or special genetic variations, but simply a
higher frequency of the disease.

Second, there are some less common genetically-linked diseases that
are found primarily in particular ethnic groups. Tay-Sachs disease, for
example, is found primarily in Jews of Eastern European origin, although
it is also found at an unusually high level in certain extended families
in Quebec and at a very low level in the general population. Some indigenous populations suffer from such unusual diseases. Research on
those diseases may lead to improved abilities to predict and treat them,
which could directly benefit the population involved. Ironically, commercialization raises a quite different problem for such diseases. Drug
companies want to maximize their profits; a test or a treatment that primarily affects a small or impoverished community may well have little
commercial value.

Third, there is at least one more speculative possibility for commercial
value. It is conceivable that some families may have genetic characteristics that are unusually beneficial and that could be developed in a way
to benefit all people. Two examples often cited are an extended family
from a village in Northern Italy who may be protected against some
forms of heart disease and a group of prostitutes in Nairobi who do not
become infected with HIV, in spite of repeated exposure. Whether either
of these examples turns out to be scientifically correct, let alone commercially valuable, remains to be seen. There seems to be no precedent
for the development of a product from a particularly "healthy" human
genetic variation, but it cannot be called impossible.

The Problem of Possible Exploitation

There is no gold rush to find commercially valuable genetic sequences among indigenous peoples and no reason to think that there will be. But it is possible that some research with indigenous peoples will have commercial value. What then?

Much of the discussion, by Rural Advancement Foundation International (RAFI) and others, has focused on patents. Life-related patents raise a host of tangled issues. The U.S. and some other countries have issued patents on cell-lines (as in the Papua New Guinea case), on genes, and on living organisms. People have attacked such patents from many different perspectives. Some have raised religious issues; others have expressed concerns about life-related patents and the environment, exploitation of individuals, unfairness to the developing world, the propriety of patenting pharmaceuticals, and whether the gene itself is the appropriate stage in the drug development process for patent protection.

I share some of these concerns. The way the patent system is currently applied to biologically-derived products seems illogical, inefficient, and sometimes offensive. Personally, I would prefer a different system, one that, like the U.S. Orphan Drug Act, conferred not a patent but a limited period of regulatory protection for a new therapeutic product. But, fundamentally, patents are a diversion from the real question: how can the rights of indigenous peoples be protected? Patents can exploit or, as with the Hagahai, protect indigenous peoples. Exploitation, or protection, can also occur wholly outside the patent system. In the famous case of John Moore (not Professor John Moore writing for this issue), the University of California holds a patent on a cell-line derived from his blood cells, but, whatever the merits of that case, the patent has proven irrelevant: the university has neither licensed nor enforced the patent in the twelve years since it was granted.

People advocate different approaches to protecting the rights of indigenous peoples in the rare cases where genetic research with them leads to commercially valuable products. This section describes five approaches, with their advantages and their problems: a ban on life-related patents, treatment along the lines of the Biodiversity Convention, gifts, individual property rights, and group rights.

One widely proposed solution, endorsed in general terms by RAFI, is to ban some or all kinds of life-related patents. This would prevent people, indigenous or not, from being "commodified" or "owned" by others, which may have great symbolic importance. But any other benefits for indigenous peoples are difficult to see. Banning patents neither guarantees any financial return to indigenous populations, nor prevents financial returns to others from inventions derived from indigenous populations.

Rather than patenting cell-lines or genes derived from particular people, companies presumably would have to reach a further step in the process, such as the production of an effective test or drug, in order to get patent protection. But whether the patent is on the gene or on the drug, there is no guarantee that the people who provided the genetic material will benefit. Drug companies could continue to profit while the populations that participated in the research that led to the drugs would continue to get nothing. And, if the ban eliminated the possibility of some kind of property-like protection for a drug company, it might also result in the loss of at least some new treatments that could benefit all of the world's peoples.

But even if such a patent ban is a good idea, it would not happen fast. The agreement of a broad spectrum of nations would be necessary to make patenting uneconomical—in light of their dominant positions in the international pharmaceuticals market, the United States, Japan, and the European Union would all have to ban such patents to make them impracticable. Such an agreement may never come; even if it does come, it does not answer the question of what should be done during the lengthy meantime.

A second alternative is to adopt the approach taken by the Biodiversity Convention. This multilateral treaty established that national governments have both rights and duties with respect to most life-forms within their borders and their genetic sequences. The national governments undertook duties to protect biodiversity and to make their local microbial, plant, and animal species available for research and commercial use. In return, the treaty requires that national governments receive just compensation for such commercial use.

This alternative seems the worst possible outcome for indigenous peoples. It would recognize their genes as effectively the property not of themselves, but of their national governments. That seems offensive in general; it is particularly chilling in light of the history of conflict between national governments and indigenous populations. Surely, no Native American tribe would be eager to see the United States government receive the royalties from the commercial use of the DNA of tribe members.

A third approach focuses on the relationship between the people who participate in the research and the researchers. It insists that samples must be freely offered gifts in order to avoid turning human materials into "property" or "commodities." Implicit in this view is the idea of full, informed consent in that the research subject makes a conscious and knowing gift of biological materials in the expectation not of personal gain, but of improvement for all humanity. The donor should be motivated not just by altruism but by a sense of reciprocity—she benefits from the gifts of myriad predecessors and passes on, by her gift, a benefit to future generations. This philosophy underlies much regulation throughout the world of the medical use of blood or organs.

This seems an admirable approach in general and it has the great virtue of focusing on the relationship between the research participants and the researchers. But it fits poorly with the situation of most indigenous peoples. It depends strongly on the idea of reciprocal advantage, but many indigenous peoples, inside and outside the "developed" world, have not received the complete benefits—or sometimes any benefits—of modern medicine and science. Research participants who are full political and economic members of "developed" societies might be expected to feel bound by such reciprocity; most indigenous peoples should not.

A fourth approach also focuses on the relationship between the researcher and the individual research participant, but in a very different way. It stresses the individual property rights of the research subject. Let each person negotiate any possible deal for the commercial use of his or her biological material. Sell blood, sell kidneys, sell genes . . . let people make their own decisions as long as they do so with full information.

Application of this kind of free market approach to organs and to bodies generates some support and much outrage. Totally apart from the strong general objections to this approach, it faces several practical problems in genetic research. First, it is hard to imagine the process that could lead every individual to become sufficiently informed about genetics, the biotechnology industry, and economics to strike a truly informed bargain for the use of their materials. That is true with any individuals; it will likely be particularly true with communities far removed from the industrialized world's science, economics, and business. Second, it will rarely, if ever, be the case that any one person will be the right person with whom to bargain. Success in finding genes requires many genetic samples. Negative samples are often as important as positive ones in pinpointing a gene's location. And positive samples are likely to come from numerous members of affected families. Which of the many individual research participants "sells the gene" to the researchers? And what keeps the researchers from playing off research participants against each other, in search of the least expensive source?

Group rights provide a fifth option, the one I support. In this approach, the decision whether and on what terms a population will participate in genetic research belongs to the group. This is the approach being suggested by the North American Regional Committee of the Human Genome Diversity Project. Its draft Model Ethical Protocol for the Collection of DNA Samples would require researchers to obtain informed consent to the research from both the participating individuals and the community to which they belong. The community would also have to be asked how it wants to handle any possible commercial value from the research. Alternatives might include requiring subsequent negotiations with the community before any commercial use, setting compensation for commercial use through a percentage royalty fixed in advance, or any other system the community chooses. The community's wishes would be enforced by contractual obligations imposed on anyone who seeks to use the material or data obtained from them.

This solution is not perfect. It requires individual assessment of important issues like deciding when group treatment is appropriate, as it clearly would be for Native American tribes but would not be, perhaps, for Irish-Americans. It requires decisions about who is "the community"—an extended family, a local community, a tribal government or a larger language or cultural group. It demands decisions about who can make legitimate decisions for the group and how those decisions are properly reached. And it requires diligent enforcement, both of the requirement that the researchers obtain the necessary group consent and decision-making and that the group's wishes are followed.

But, unlike any of the other solutions, it offers indigenous communities an attractive set of possibilities. It establishes that the communities being researched will have realistic control over the research and its commercial uses. The community could decide to allow patents or to ban patents, to prohibit commercialization or to benefit from it, to participate in the research or not participate in it. The group nature of the decision fits well with both the realities of many indigenous communities and with the realities of genetics research, where the entire population studied contributes to the result, not just the individuals from whose samples a sequence of interest is isolated.

References

Risch, Neil & David Botstein. 1996. A Manic Depressive History, *Nat. Gen.* 12: 351.

Source: Henry T. Greely, "Genes, Patents, and Indigenous Peoples: Biomedical Research and Indigenous Peoples' Rights," *Cultural Survival Quarterly*, summer 1996, 54–57.

DOCUMENT 85: A. Rural Advancement Foundation International, "Indigenous Person from Papua New Guinea Claimed in US Government Patent" (1995)

Others are opposed to the genetic analysis of indigenous peoples, as evident in the following documents from the Rural Advancement Foundation International.

* * *

While the rest of the world is seeking to protect the knowledge and resources of indigenous people, the National Institutes of Health (NIH) is patenting them. "This patent is another major step down the road to the commodification of life. In the days of colonialism, researchers went after indigenous people's resources and studied their social organizations and customs. But now, in biocolonial times, they are going after the people themselves," says Pat Roy Mooney, RAFI's Executive Director. . . .

. . . RAFI believes that this is only the beginning of a dangerous trend toward the commodification of humanity and the knowledge of indigenous people. Whether human genetic material or medicinal plants are the target, there is scarcely a remote rural group in the world that is not being visited by predatory researchers.

Source: Rural Advancement Foundation International, "Indigenous Person from Papua New Guinea Claimed in US Government Patent" (Press release, October 5, 1995).

DOCUMENT 86: B. Statement by Abadio Green Stocel, President of National Indigenous People's Organization of Colombia (1996)

Thousands of samples of DNA from dozens of Colombian indigenous and rural peoples have been exported to the U.S. without our knowledge. HGDP (Human Genome Diversity Project) may have benign intentions; but we will not allow it to proceed in our communities until we are absolutely sure the project isn't going to take control of our genetic heritage and put it in the hands of Northern lawyers, technocrats, and biotechnology companies.

DOCUMENT 87: C. Statement by Ruth Liloqula, Citizen of the Solomon Islands (1996)

The project has very little interest in helping indigenous people to survive, or in addressing the social, the economical, the political and the exploitation issues that engage these indigenous groups of people. . . . For Solomon Islanders, however, the traditions and customs that protect the sanctity of human materials—be it blood, cloth, saliva, or mucus— are very strong. It pays no respect to our dignity, culture and identity—

and at the same time it will eventually turn our people away from the medical help they need.

Source: Rural Advancement Foundation International, "Diverse Group Joins in Washington to Oppose U.S. Government in Funding for the HGDP" (Press release, September 11, 1996).

DOCUMENT 88: David F. Betsch, "DNA Fingerprinting in Human Health and Society" (1994)

Another development in genetics is DNA fingerprinting. This document gives a definition of this application.

* * *

Like the fingerprints that came into use by detectives and police labs during the 1930s, each person has a unique DNA fingerprint. Unlike a conventional fingerprint that occurs only on the fingertips and can be altered by surgery, a DNA fingerprint is the same for every cell, tissue, and organ of a person. It cannot be altered by any known treatment. Consequently, DNA fingerprinting is rapidly becoming the primary method for identifying and distinguishing among individual human beings.

Source: David F. Betsch, "DNA Fingerprinting in Human Health and Society" (Biotechnology Training Programs, Inc., Iowa State University Office of Biotechnology, 1994).

DOCUMENT 89: Michelle Eadie, "Science on Trial"

As with many technologies, DNA fingerprinting was developed to solve one problem but was then applied to other areas.

* * *

DNA fingerprinting/profiling was developed in 1985 by Alec Jeffreys and colleagues at Leicester University [England]. At this stage it was mainly being used for paternity testing in cases of doubtful parentage and to screen for genetically inherited diseases, but already its applicability to forensic science for identification of criminals, from traces of blood stains, semen stains, and hair roots left at the scene of crimes, was being discussed. In 1986 DNA fingerprinting became commercial for use in paternity testing, and potential problems, such as interpretation of results in forensics, were speculated.

In 1989 there was news of the first DNA fingerprinting evidence to be used in trials in Australia, although they had been used in the United States and England since 1987.

Source: Michelle Eadie, "Science on Trial"; found at Anzwers Search Centre. http://www.ozemain.com.au/~dtebbut/oj/ojsci3.html.

DOCUMENT 90: James L. Mudd, "Quality Control in DNA Typing: A Proposed Protocol" (1988)

DNA testing can be used in crime investigation and in determining identity and relatedness of people. This type of research has proved helpful to the Federal Bureau of Investigation.

* * *

The implementation of DNA typing in the FBI Laboratory marks an important advancement in forensic science. Because of the complexity of the technique and the impact of the data, it is essential that the procedure be scientifically sound and that the quality of the analytical data generated be maintained. These goals can be met, through implementation of a detained Quality Assurance/Quality Control program. This program consists of separate QA and QC functions. The QA is an external function consisting of blind and open proficiency test samples and is used to verify that the QC function is being performed adequately. The QC program is an internal function that ensures that the DNA typing results satisfy established performance criteria.

The combined QA/QC program not only monitors the DNA typing process, but also provides laboratory managers and examiners, as well as the judicial system, with the confidence that the quality of the analytical process and accurate interpretation of the data are maintained.

Source: James L. Mudd, "Quality Control in DNA Typing: A Proposed Protocol," *Crime Laboratory Digest* 15 (1988): 105–111.

DOCUMENT 91: Gina Kolata, "The Code: DNA and O. J. Simpson: Testing Science and Justice" (1994)

DNA research has led to various other applications.

* * *

The speculation now among the hordes of lawyers who are microscopically examining the pretrial maneuvering is that the case may hinge on the delicate science of DNA fingerprinting and all the controversy that surrounds it.

In theory, nothing is more individual than a person's genetic code. A DNA test can look at the genes in a drop of blood or a strand of hair and judge whether they came from Mr. Simpson or from someone else. . . . Done correctly, the test could establish Mr. Simpson's innocence or convince a jury that he is guilty.

But the scene of a crime is not subject to the rigid controls of the laboratory. And judges and jurors are generally not molecular geneticists. With all the uncertainties over how to interpret the data, there is always the danger that justice will be led astray, or confused by the dueling opinions of expert witnesses.

Source: Gina Kolata, "The Code: DNA and O. J. Simpson: Testing Science and Justice," *New York Times*, June 26, 1994, sec. 4, p. 1.

DOCUMENT 92: *USA Today*, "DNA Likely Was Contaminated, Simpson Defense Asserts" (1996)

Videotape of a police technician haphazardly collecting blood samples drew gasps at O. J. Simpson's civil trial Thursday, narrated by a defense expert who said the DNA evidence was contaminated and worthless.

Microbiologist John Gerdes conceded under cross-examination, however, that with the exception of a couple of samples, there was "no direct evidence" of contamination, only of risky collection techniques and unusual results.

Defense attorney Robert Blasier, questioning Gerdes, pressed his point that sophisticated DNA results from two highly respected labs were worthless because they were based on such contaminated samples.

In cross-examination, however, Gerdes acknowledged that just because samples could be contaminated didn't mean that they were.

Source: USA Today, "DNA Likely Was Contaminated, Simpson Defense Asserts,"
December 12, 1996.

DOCUMENT 93: Joe Jackson, "DNA Evidence on Trial Again in Virginia" (1995)

The blood-splattered clothing that sent Joseph Roger O'Dell III to death row nearly a decade ago now could be his strongest bid for freedom.

On Tuesday, O'Dell's lawyers will argue in federal appeals court in Richmond that he deserves a new hearing because new DNA evidence casts doubt on his conviction for the February 1985 rape and murder of Helen Schartner in Virginia Beach.

Once again, DNA is on trial in Virginia—the first state to develop DNA fingerprinting as a crime-fighting tool and the first in which a man was executed due to DNA evidence. If anything, the O'Dell case has become a mirror of the legal warfare that arises when DNA enters the scene.

Lawyers could debate the result [of the tests], but one thing was certain. The court had ruled that the only testable DNA yielded results different from the trial evidence—the same evidence that convinced jurors that Joseph O'Dell must die.

Source: Joe Jackson, "DNA Evidence on Trial Again in Virginia," *Virginia Pilot*
[Norfolk], December 4, 1995.

DOCUMENT 94: Robert Boyd, "DNA on File for Millions of Americans" (1994)

Of course, DNA research raises a variety of ethical concerns.

* * *

Thanks to the O. J. Simpson case, just about everyone knows a substance called DNA can identify a criminal suspect with unusual accuracy.

But most people don't know that a nationwide system of DNA databanks is rapidly being built. It already contains blood samples from millions of Americans—whether or not they have been involved in a crime.

Someday, these growing collections might cover every citizen, helping to solve murders, find missing children, identify plane-crash victims or free an innocent man on death row.

However, some scientists, doctors and lawyers fear an unprecedented threat to individual privacy.

"DNA profiling poses a special risk of invasion of privacy concerning personal and medical traits," the National Research Council, a high-level scientific advisory committee, declared in 1992. "Such information could lead to discrimination by insurance companies, employers or others against people with particular traits."

Source: Robert Boyd, "DNA on File for Millions of Americans," *Seattle Times*, November 1, 1994, p. A1.

DOCUMENT 95: "HUGO (Human Genome Organization—Europe) Ethics Committee, Statement on DNA Sampling" (1998)

The family is the nexus of a variety of relationships (legal, moral, social and biological). Irrespective of legal definitions of the family and of its different social and cultural configurations, genetic research may yield genetic information that is important to immediate relatives. The very fact of participation in research or not, or the decision to refuse to warn at-risk relatives to withdraw, or the failure to provide for access after death, all affect the interests of present and future relatives.

These shared biological risks create special interests and moral obligation with respect to access, storage and destruction that may occasionally outweigh individual wishes. A different response is necessary, however, in relation to institutional third parties, such as employers, insurers, schools, and government agencies, because of possible discrimination. Counselling prior to participation is also necessary to avoid stigmatization. Standardization of procedures and the security of samples are essential.

Source: "HUGO (Human Genome Organization–Europe) Ethics Committee Statement on DNA Sampling: Control and Access," *Eubios Journal of Asian and International Bioethics* 8 (1998): 56–57.

DOCUMENT 96: Mary-Claire King, "An Application of DNA Sequencing to a Human Rights Problem" (1991)

This document shows the practical relevance of DNA technology when used to identify the family of origin of children who were kidnapped during seven years of military rule in Argentina. Of critical importance are the technical developments needed to link these children with their grandparents, since frequently their parents were murdered or died during torture. Equally important is trying to determine the placement of the children after their identity is discovered.

* * *

1. INTRODUCTION: THE HISTORICAL CONTEXT

Between March 1976 and December 1983, the Republic of Argentina was controlled by a military junta responsible for the abduction, torture, and murder of thousands of citizens (Amnesty International, 1980; Interamerican Commission on Human Rights, 1980). The total number of victims of extra judicial execution will probably never be known, but in December 1983, the newly elected democratic government established the National Commission on the Disappearance of Persons (CONADEP), with the request that this commission document, insofar as possible, the history of disappearances of citizens during the period of military rule. By September 1984, CONADEP had prepared evidence on 8800 victims; however, CONADEP and human rights organizations believe the actual number of "disappeared" persons to be much higher (National Commission on the Disappearance of Persons, 1984).

Approaches from human genetics have been used to help identify a special subset of these victims: the 210 children who were kidnapped at birth or as infants by military and police who murdered their parents and retained or sold the children. Unlike the older victims of the "Dirty War," many of these very young children remained alive, but were made to "disappear" (Abuelas de Plaza de Mayo, 1984). In 1977, during the period of military rule, the surviving relatives of kidnapped children formed the Grandmothers of the Plaza de Mayo, a human rights group devoted to finding kidnapped children and reuniting families. Through persistent collection and follow-up of circumstantial evidence, the Grandmothers began to locate their kidnapped grandchildren, primarily in the households of military and police officials and their collaborators (Nosiglia, 1985).

It soon became apparent to the Grandmothers that it was necessary, but not sufficient, to establish that a specific child was a kidnap victim. It was also necessary to establish each child's true identity by objective means. After the fall of the military and the election of a democratic government, it became possible to bring charges in Argentinian courts against the kidnappers. The Grandmothers therefore asked geneticists for help in establishing objectively the identities of these grandchildren. In June 1984, at the request of the Grandmothers of the Plaza de Mayo and CONADEP, a commission from the American Association for the Advancement of Science first traveled to Buenos Aires to help develop procedures for genetic identification of kidnapped children.

In 1984, genetic relationships were most effectively established using serological testing of human lymphocyte antigens (HLAs) at the A, B, C, and DR loci (Albert et al., 1984). The immunogenetics laboratory of Ana Maria DiLonardo at the Durand Hospital in Buenos Aires, which was experienced in HLA serologic typing, agreed to undertake the identification project. That laboratory has been successful in establishing the identities of a large number of children and in presenting evidence to the Argentinian courts that has led to the reunification of families (DiLonardo et al., 1984; Diamond, 1987).

In the intervening 7 years, the genetic approaches applied to establishing the identity of these kidnapped children have been extended. Two developments motivated this work. First, at the request of the Grandmothers and other human rights organizations, the government of Argentina established in 1985 the National Genetic Data Bank. This voluntary service offers grandparents, aunts, uncles, cousins, and other surviving relatives of disappeared children the opportunity to have their blood sampled, HLA serology determined, and white cells stored for extraction of DNA. Pedigrees documenting the relationships of surviving relatives to the murdered parents and missing children are also constructed. The intent of the legislation is that, as kidnap victims are discovered, their identities can be determined by matching the child to surviving relatives. Several hundred surviving relatives have contributed to the Data Bank in the past 7 years. Consequently, as kidnap victims are found for whom little circumstantial evidence of identity exists, each must be matched against multiple families. To establish identity with a sufficiently high level of statistical certainty under these circumstances usually requires matching to more than one genetic linkage group.

The second development during the project has been the appearance of families with few surviving relatives. Deaths from natural causes and by murder during the military period and fear have led to incomplete information for some families. It is not uncommon for a family searching for a missing child to include only the maternal lineage, sometimes only the mother or a sister or brother of the murdered mother of a kidnap victim. Therefore, it has been necessary to employ a genetic approach that permits the identity of a child to be determined even if only a single maternal relative is alive.

The major concern underlying this work is to do what is best for the children (Abuelas de Plaza de Mayo, 1984). Clearly, circumstances vary enormously and the custody of each child must be decided individually.

A few of the children born in captivity or kidnapped as infants were adopted in good faith by families with no ties to the military. The resolutions of these exceptional cases have generally been amicable, with the children being told the truth about their biological parents and spending time with both their biological and adoptive families. However, the cases of children living with military or police officers involved in the torture and murder of their parents are far more difficult. These have comprised the vast majority of children discovered so far. Certainly under normal circumstances, a child would not be left with kidnappers or their accomplices regardless of his or her age at abduction. The notion of assessing whether persons involved in kidnapping, torture, or murder are suitable parents for the children of their victims appears highly unlikely. Kidnapping has universally been considered a crime. Is the situation different in Argentina because kidnapping occurred on a large scale? The human rights groups with whom we work suggest that to abandon the search for the kidnapped children of Argentina is to abandon a group of children who will not grow up in carefree innocence. As these children become adults, what would their attitudes be toward relatives who knew they had disappeared but did nothing? What would be the effect on a young person to learn he or she has lived with people involved in the murders of his/her parents and that his/her surviving relatives did nothing to find him/her? Would failing to attempt to identify the kidnapped children implicitly grant immunity to kidnappers? Would this increase the sense of invulnerability of abusers of human rights in other countries?

The historical situation that led to this application of genetics to human rights is unprecedented. Thus, answers to these questions of ethics, law, and mental health are developing with our current experience. Meanwhile, as of this time (early 1991), the political situation in Argentina is much more hostile to the Grandmothers' efforts than during the Alfonsin presidency of 1984–1989. In particular, it is increasingly difficult to work within the Argentinian judicial system. However, the Grandmothers remain undaunted. They point out that the average age of the kidnapped children is now 15 years. Very soon, these children will have the legal right to determine for themselves their identities. For this purpose mtDNA sequences will be available. Even though the grandparents of a kidnaping victim may die before the grandchild is found, the young adult's maternal lineage will be identifiable using the genetic information the Grandmothers have left behind. A young person can thereby be put in touch with his/her family—surviving aunts, uncles, and cousins— and his/her history. For the past 15 years, the Grandmothers have been searching for their kidnapped grandchildren. Very soon, these grandchildren will come looking for them.

References

Abuelas de Plaza de Mayo. (1984). "Ninos Desaparecidos: Su Restitucion. Consclusiones del Seminario National." Buenos Aires.

Albert, E. D., Baur, M. P., and Mayr, W. R. (1984). "Histocompatibility Testing 1984." Springer-Verlag, New York.

Amnesty International. (1980). "Political Killings by Governments." Amnesty International Publications, New York.

Diamond, J. M. (1987). Abducted orphans identified by grandpaternity testing. *Nature* 327: 552–553.

DiLonardo, A. M., Darlu, P., Baur, M., Orrego, C., and King, M. C. (1984). Human genetics and human rights: Identifying the families of kidnapped children. *Am. J. Foren. Med. Pathol.* 5: 339–347.

Interamerican Commission on Human Rights. (1980). "Report on the Situation of Human Rights in Argentina." General Secretariat, Organization of American States, Washington, D.C.

National Commission on the Disappearance of Persons. (1984). "Nunca Mas." Editorial, Universitaria de Buenos Aires, English translation: 1986, Faber and Faber Ltd., London.

Nosiglia, J. E. (1985). "Botin de Guerra." Cooperativa Tierra Fertil, Buenos Aires.

Source: Mary-Claire King, "An Application of DNA Sequencing to a Human Rights Problem," *Molecular Genetic Medicine* 1 (1991): 117–131.

DOCUMENT 97: George Annas, "Who's Afraid of the Human Genome?" (1989)

Here George Annas sums up the volatility of the future of the use of the knowledge we've gained through genetic research, expressing a sentiment most likely echoed by others in the field.

* * *

Although we are utterly unprepared to deal with issues of mandatory screening, confidentiality, privacy, and discrimination, we will likely tell ourselves that we have already dealt with them well.

Source: George Annas, "Who's Afraid of the Human Genome?" *The Hastings Center Report* 19 (July–August 1989): 19–21.

Part VI

Diagnostic Applications of Genetic Information

Scientists hope many of the developments in genetics will provide a very practical payoff in the use of this knowledge and various techniques to provide therapy for disease. The Human Genome Project was essentially funded on the basis of this hope, and much genetic research focuses on both diagnostic and therapeutic application. This section presents several documents that discuss various therapies and their outcomes. Generally, although progress is being made, the progress is slow and many outcomes are uncertain. As with other complex processes, progress will be slow. However, developments thus far have encouraged a variety of initiatives.

DOCUMENT 98: Richard C. Mulligan, "The Basic Science of Gene Therapy" (1993)

Human gene therapy relies upon a variety of techniques that must be coordinated so that specific genes can be delivered to the appropriate target genes. A description of several of these techniques are explained in the following documents.

* * *

Ever since the development of recombinant DNA technology, the promise of the technology for dramatically improving the practice of medicine has been vigorously championed. . . . Recombinant DNA technology has also produced the means for defining the roles of specific gene products in the pathogenesis of human disease. . . . [R]esearch has indicated that successful implementation of gene transfer in the clinic will require the coordinated development of a variety of new technologies and the establishment of unique interactions between investigators from divergent medical and basic science disciplines.

The transduction of appropriate target cells represents the critical first step in gene therapy; consequently, the development of methods of gene transfer suitable for different forms of therapy has been a major focus of research. The single common feature of these methods is the efficient delivery of genes into cells.

Source: Richard C. Mulligan, "The Basic Science of Gene Therapy," *Science* 260 (May 14, 1993): 928.

DOCUMENT 99: Marcia Barinaga, "Gene Therapy for Clogged Arteries Passes Test in Pigs" (1994)

Cardiologists have known for a long time that abnormal growth of the smooth muscle cells lining artery walls plays a key role in the blockage of arteries during coronary artery disease. It also contributes to reblockage [restenosis] of many arteries that have been opened by balloon angioplasty or replaced in bypass operation.

The Nabel team first performed balloon angioplasty on the groin arteries of pigs, an insult that triggers smooth muscle cell proliferation in an otherwise healthy artery. Then they used a modified form of a human respiratory virus called adenovirus to carry the tk gene into the damaged arterial cells.

The tk enzyme doesn't harm the proliferating cells directly, but it makes them susceptible to ganciclovir. This drug, which is also harmless by itself, is phosphorylated by the tk enzyme, causing it to insinuate itself into the DNA of dividing cells and halt cell division. . . . [C]ompared to control animals, the pigs treated with the tk gene and ganciclovir had a 50% to 90% reduction in artery wall thickening, specifically in the area where smooth muscle cell proliferation occurs.

Source: Marcia Barinaga, "Gene Therapy for Clogged Arteries Passes Test in Pigs," *Science* 265 (August 5, 1994): 738.

DOCUMENT 100: S. Rosenberg and others, "Use of Tumor-Infiltrating Lymphocytes and Interleukin-2 in the Immunotherapy of Patients with Metastatic Melanoma" (1988)

As development in the field of genetic engineering has progressed, researchers have sought ways to treat cancers in humans. This document reports on a treatment of malignant melanoma (skin cancer). This protocol calls for the isolation of immune cells (T cells) from tumors taken out of human subjects, then grown to large numbers in tissue cultures with the T cell growth factor interleukin-2 (IL-2). The larger T cells are then injected into the patients along with high doses of IL-2 as their treatment. This procedure has proven to be expensive and clinically difficult.

* * *

Conventional chemotherapy is relatively ineffective in the treatment of patients with metastatic melanoma, and approximately 6000 patients die of this disease in the United States each year. We have sought new immunotherapeutic approaches to the treatment of patients with melanoma and have reported that regression of metastatic melanoma could be achieved in some patients treated with high doses of interleukin-2 alone or in combination with the adoptive transfer of LAK cells. A recent analysis of our experience has revealed that objective evidence of cancer remission was observed in 10 of 48 patients. . . .

This treatment should be considered highly experimental and should be pursued in centers where the immunologic factors required for successful treatment can be assessed. The present study does demonstrate, however, that the adoptive transfer of immune atologous cells can be effective in mediating cancer regression in selected patients, and further emphasizes the need to pursue the development of this biologic approach to cancer therapy.

Source: S. Rosenberg and others, "Use of Tumor-Infiltrating Lymphocytes and Interleukin-2 in the Immunotherapy of Patients with Metastatic Melanoma," *New England Journal of Medicine* 319 (December 22, 1988): 1676–1680.

DOCUMENT 101: Natalie Angier, "Gene Implant Therapy Is Backed for Children with Rare Disease" (1990)

The first major clinical trial gene therapy was used on a young girl with ADA deficiency—the disease that in effect shuts down one's immune system. The cure to this point had been to enclose patients in a totally sterile environment with complete isolation from all direct human contact. This experiment introduced a billion cells, each with a foreign gene in it, to help her body produce the enzyme AMA that she lacks. The research was a success.

* * *

In their trials Dr. Blaese and Dr. Anderson plan to correct ADA deficiency by taking blood from the patient, isolating the T-cells, growing them in vast quantities in the laboratory, and infecting them with a genetically engineered virus.

The virus is a mouse retrovirus that can infect human cells but has been carefully redesigned so it cannot replicate or cause disease in humans. The researchers have sliced into the crippled mouse virus a copy of the human gene for the ADA enzyme.

Once the viral package has infected the patient's cultured T-cells, it can integrate into the chromosomes, where the gene for ADA will switch on and begin producing the life-saving enzyme. The researchers will then inject the T-cells back into the patient through a simple blood transfusion.

Source: Natalie Angier, "Gene Implant Therapy Is Backed for Children with Rare Disease," *New York Times*, March 8, 1990, A1 and B9.

DOCUMENT 102: W. French Anderson, "Human Gene Therapy" (1992)

One of the anticipated benefits of the Human Genome Project is the capacity to do gene therapy as a way to treat disease. Various types of gene therapy protocols are being examined. Several safety and ethical issues are an inherent part of this therapeutic breakthrough. Many argue, as does the author of the following document, that the technology should be limited to the treatment of disease, not applied in various forms of enhancement therapy.

* * *

Safety Considerations

How safe are these protocols? The safety considerations in retroviral-mediated gene transfer and therapy clinical protocols have been extensively reviewed (1). There is now experience from the equivalent of 106 monkey-years and 23 patient-years in individuals who have undergone retroviral-mediated gene transfer. Side effects from the gene transfer have not been observed, pathology as a result of gene transfer has not been found, and there has never been a malignancy observed as a result of a replication-defective retroviral vector.

Recently, however, investigators at NIH described three monkeys who developed malignant T cell lymphomas after a bone marrow transplantation and gene transfer protocol with a helper virus–contaminated retroviral vector preparation (2). The helper virus was probably directly responsible for these lymphomas. This finding strongly reaffirms the necessity for clinical protocols to use helper virus–free vector preparations, as is required for all protocols approved by RAC and by FDA.

Ethical and Social Implications

The ethical and social implications of somatic cell gene therapy have been discussed (3), and there is now a general consensus that somatic cell gene therapy for the purpose of treating a serious disease is an ethical therapeutic option. However, considerable controversy exists as to whether or not germline gene therapy would be ethical (4). The issues are both medical and philosophical.

The medical concern is that genetic manipulation of the germline could produce damage in future generations. Medicine is an inexact science; we still understand very little about how the human body works. Well-intentioned efforts at treatment with standard therapeutics can produce unexpected problems months or years later. Altering the genetic information in a patient's cells may result in long-term side effects that are unpredictable at present. Until the time comes that it is possible to correct the defective gene itself by homologous recombination (rather than just inserting a normal copy of the gene elsewhere in the genome), the danger exists of producing a germline mutagenic event when the "normal" gene is inserted. Therefore, considerable experience with germline manipulation in animals, as well as with somatic cell gene therapy in humans, should be obtained before considering human germline therapy.

Besides the medical arguments, there are a number of philosophical, ethical, and theological concerns. For instance, do infants have the right to inherit an unmanipulated genome, does the concept of informed consent have any validity for patients who do not yet exist, and at what point do we cross the line into "playing God"? The feeling of many observers is that germline gene therapy should not be considered until much more is learned from somatic cell gene therapy, until animal studies demonstrate the safety and reliability of any proposed procedure, and until the public has been educated as to the implications of the procedure. The NIH RAC plans to initiate the public debate.

There is also considerable concern about using gene transfer to insert genes into humans for the purposes of enhancement, that is, to try to "improve" desired characteristics (5). The same medical issues arise: we have too little understanding of what normal function is to attempt to improve on what we think is "normal." Correction of a genetic defect that causes serious illness is one thing, but to try to alter a characteristic such as size (by administration of a growth hormone gene to a normal child, for example) is quite another. This area is further clouded by major social implications as well as by the problem of how to define when a given gene is being used for treatment (or for preventing disease) and when it is being used for "enhancement."

REFERENCES AND NOTES

1. H. M. Temin, *J. Med. Virol.* 13, 13 (1990); *Hum. Gene Ther.* 1, 111 (1990): K. Cornetta, R. A. Morgan, W. F. Anderson, ibid. 2, 5 (1991).

2. R. Kolberg, *J. NIH Res.* 4, 43 (1992); A. Nienhuis, unpublished data.

3. J. C. Fletcher, *J. Med. Philos.* 10, 293 (1985); L. Walters, *Nature* 320, 225 (1986); *Hum. Gene Ther.* 2, 115 (1991); L. Kass, in *The New Genetics and the Future of Man*, M. Hamilton, Ed. (Eerdmans, Grand Rapids, MI, 1972), p. 61; P. Ramsey, ibid., p. 175: J. Fletcher, *The Ethics of Genetic Control* (Doubleday, Garden City, NY, 1974), pp. 14–15; President's Commission for the Study of Ethical Problems in Medicine and Biomedical and Behavioral Research, *Splicing Life* (Government Printing Office, Washington, DC, 1982).

4. J. C. Fletcher, *Virginia Law Review* 69, 515 (1983); W. F. Anderson, *J. Med. Philos.* 10, 275 (1985); A. M. Capron, in *Genetics and the Law III*, A. Milunsky and G. J. Annas, Eds. (Plenum, New York, 1985), p. 24; G. Fowler, E. T. Juengst, B. K. Zimmerman, *Theor. Med.* 10, 151 (1989); C. A. Tauer, *Hum. Gene Ther.* 1, 411 (1990); E. T. Juengst, *J. Med. Philos.* 16, 587 (1991); B. K. Zimmerman, ibid., p. 593; K. Nolan, ibid., p. 613; M. Lappe, ibid., p. 621; R. Mosely, ibid., p. 641; A. Mauron and J. Thevoz, ibid., p. 648; E. M. Berger and B. M. Gert, ibid., p. 667.

5. J. C. Fletcher and W. F. Anderson, in *Law, Medicine, and Health Care*, M. Kaitz, Ed. (American Society of Law and Medicine, Boston), in press. W. F. Anderson, *J. Med. Philos.* 14, 681 (1989); W. F. Anderson, *Hastings Cent. Rep.* 1990, 21 (January–February 1990); B. Hoose, *Hum. Gene Ther.* 1, 299 (1990); J. Porter, ibid., p. 419.

Source: W. French Anderson, "Human Gene Therapy," *Science* 256 (May 8, 1992): 808–813.

DOCUMENT 103: Jeffrey M. Isner and others, "Arterial Gene Transfer of Therapeutic Angiogenesis in Patients with Peripheral Artery Disease" (1996)

This document presents the use of genetically engineered human growth factor material to help restimulate artery growth. This allows relief for patients previously untreatable.

* * *

The age-adjusted prevalence of peripheral arterial disease (PAD) in the U.S. population has been estimated to approach 12%. The clinical consequences of occulusive peripheral disease include pain on walking, pain at rest, and loss of tissue integrity in the distal limbs; the latter may ultimately lead to amputation of a portion of the lower extremity....

The purpose of this clinical protocol is to document the safety of therapeutic angiogenesis achieved in this case by percutaneous cather-based delivery of the gene encoding vascular endothelial growth factor (VEGF) in patients with PAD; and as secondary objectives, investigate the bioactivity of this strategy to relieve rest pain and heal eschemic ulcers of the lower extremities.... In brief, a single intra-arterial bolus of VEGF recombinant human protein, delivered percutaneously to the ischemic limb via an intravascular cather, resulted in angiographic, hemodynamic, physiologic, and histologic evidence of augmented colateral artery development.

Source: Jeffrey M. Isner and others, "Arterial Gene Transfer of Therapeutic Angiogenesis in Patients with Peripheral Artery Disease," *Human Gene Therapy* 7 (May 20, 1996): 959–988.

DOCUMENT 104: Catherine V. Hayes, "Genetic Testing for Huntington's Disease—A Family Issue" (1992)

As diagnostic applications of genetic information evolve, our society is faced with new issues. Thus even though many diseases are heritable or "run in families," testing for a disease has been traditionally understood as a private or individual event because the individual being tested is the one affected by the results. Presymptomatic tests have, however, been developed for late-onset diseases such as Huntington's disease. Family members can now know in advance of the onset of the disease which members are at risk. The following document, written by the president of the Huntington's Disease Society of America, discusses several consequences of such testing. The most critical issue is a shift from viewing genetic testing as an individual issue to viewing it as a family issue. This is a major development that requires careful study.

* * *

For the first 33 years of my life, I lived at risk for Huntington's disease. Huntington's disease is transmitted in an autosomal dominant fashion: each child of a person with the disease has a 50–50 chance of inheriting the Huntington's gene. Though it may take decades for symptoms to become apparent (Huntington's disease usually strikes insidiously between the ages of 30 and 50), the gene will eventually cause the disease and lead to death after 10 to 15 years of unremitting degeneration.

As a child, I discovered the well-kept family secret that my maternal grandfather had lived out his last years in a state mental hospital suffering from Huntington's disease. I watched an aunt and uncle slowly succumb to the disease. Finally, I had to admit that my mother was also affected. My brothers and I were now, unmistakably, at risk.

Before 1983, the anxious uncertainty of those of us at risk for Huntington's disease could be brought to an end either by a positive diagnosis or by the onset of old age and the increasing likelihood that we were among the lucky ones—the "escapees." We had to wait, watch for minor twitching, mood swings, and forgetfulness, and hope that they did not become severe enough to confirm our worst fears. We knew that if Huntington's disease developed there was (and still is) nothing to be done.

One fall day in 1983, I heard on the car radio that a genetic marker, a segment of DNA believed to be close to the gene for Huntington's disease, had been discovered. It was clear to me that within a few years a presymptomatic test would be available. I had to stop the car because I was crying. I knew that I would take the test. Knowing, whatever the outcome, would be better than waiting and wondering day after day.

A presymptomatic test was indeed developed in 1986, and in the summer of 1987 I enrolled in the testing research project at Johns Hopkins University Hospital. It was a long and complex process. Blood samples from numerous members of my family had to be collected and analyzed. I underwent several months of genetic counseling to clarify my motivation and to determine my ability to cope with any possible outcome.

After a period of months, nothing remained but the nerve-racking wait for the results.

Finally, the wait was over: my test was negative. The DNA analysis had shown with 96 percent certainty (later increased to 99 percent, with refinement of the testing process) that I had not inherited the gene for Huntington's disease. When I learned the results I cried and laughed. It took months for the news to sink in. I am still adjusting.

The incomparable relief I felt at finally being free of the fear and uncertainty (a feeling shared by many who have been tested, as Wiggins et al. report in this issue of the *Journal* (1)) was tempered by the painful knowledge that other family members had not been and would not be so lucky. I have five brothers and eight nieces and nephews. One brother already has symptoms. He has two children. Another brother has been tested and, like me, received a negative result. He has married and had his first child. A third brother, also with two children, took the test and was found to have a high probability of having the Huntington's gene.

As president of the Huntington's Disease Society of America, I have learned that many medical professionals have difficulty viewing genetic issues in a family context. This is not surprising given the conventional emphasis in medicine on individuals. Moreover, the care of a person with a chronic and complex disorder such as Huntington's disease places a heavy demand on an overburdened health care system ill equipped to address secondary issues involving "unaffected" family members.

Nevertheless, a diagnosis of Huntington's disease affects every member of the family. Families and family members deal with it in various ways. Some people confront it openly, some will not talk about it, and others dwell only on the negative aspects. In short, no one can control how families will handle genetic issues, nor can anyone predict individual coping strategies. For example, there is often a nosy relative intent on piecing together a family pedigree (and other family members who insist on absolute privacy). I believe strongly that researchers should not feel compelled to do anything about such situations. Inquisitive relatives are a part of every family, and if the family can do nothing to quell their curiosity, then neither can anyone else.

Nonetheless, some general guidelines might be useful in dealing with families. First and foremost, genetic testing must be viewed as a family issue, not an individual one. The person who enrolls in a testing program should be strongly encouraged to involve other family members, within reason. Testing one member of a family will affect other members. Persons who refuse to involve their families may not have considered fully the consequences for other members or for themselves.

A case in point involves a pair of identical twins only one of whom wanted to be tested. She swore that she would never reveal the results to anyone else in the family, in particular her twin. The researchers agreed to test her. Once she was informed of the results—that there was a high probability that she would have Huntington's disease—the information spread quickly throughout the entire family. This meant that the twin who did not want to know her genetic status was now faced with the unwelcome knowledge that she too would probably have the disease. Had the researchers considered all the implications of testing, they might have referred both sisters for family counseling, providing them the opportunity to resolve the question of whether to be tested. One person's right to know in this case did not necessarily outweigh her twin's right not to know. When information was given to one twin, the other irretrievably lost the freedom to decide.

Most researchers cannot possibly know what it is like to grow up in a family haunted by a genetic disease—nor should we expect them to. But a greater effort to empathize could help improve their understanding of families and enhance the relationship between doctors and families. This point was most clearly illustrated for me during a recent symposium on genetic diseases. One participant reported that 25 percent of family members in a survey said they would terminate a pregnancy in which the baby faced a "serious medical issue," but not one in which the baby would have the same disability as the respondent. Most of the participants shook their heads in disbelief. For me and for people like me, this was no surprise. We grow up among family members affected by a genetic disorder. They are real people to us, our own flesh and blood. We are accustomed to the changes that the disease brings about, and although we may be fearful, we love and value our affected relatives. For many of us, abortion might be the best option when there is a risk of some other medical condition, but not our own.

Another difficult moral issue is that of suicide when the likelihood of having a genetic disease becomes known. It has been speculated that genetic testing may lead to an increased rate of suicide among those who test positive. To date, post-testing follow-up has not borne out this hypothesis, as Wiggins et al. note. The possibility of suicide is greater when symptoms begin to appear than in the immediate aftermath of testing. Obviously, the topic of suicide should be carefully probed in the early stages of genetic counseling. When it is discussed, counselors will generally find that they are not exploring virgin territory. Most people at risk for Huntington's disease have thought long and hard about what they might do if symptoms develop and may even admit these thoughts. Most of those who actually ask for genetic information are relatively prepared to handle the facts.

The rapidly advancing field of molecular genetics will eventually have far-reaching consequences for everyone. Although Huntington's disease continues to be a closely guarded secret in some families, for others radical changes in science are already being matched by radical changes in attitude. There are clear signs of an emergent openness about the disease. My grandfather lived out his last days in an institution, and I did not even know he was alive. My five brothers married and most had children before we openly acknowledged the existence of Huntington's disease in our family. But my 15-year-old nephew has readily stood up in class and told his fellow students about Huntington's disease (in contrast to his teacher, who in reviewing various conditions, had difficulty pronouncing "Huntington's disease"). My older brother (who has Huntington's disease) recently agreed to be featured in *People* magazine.

Of course, denial still exists. My mother, who has lived in a nursing home for years now, still tells me that she does not have "that disease." Denial is after all a valuable coping mechanism, particularly for people who have little else to fall back on. It should not be taken away until there is something better—effective treatment, a cure—to replace it. Many people who said they would be tested have changed their minds. This is understandable to those of us at risk for Huntington's disease. In the abstract, it is easy to think that you would want to know. When it is time to find out, the possibility of a positive test is frightening. How many of us would really choose to be told how we will die?

Perhaps I sound as if I am opposed to genetic testing. I am not—I took the test myself—but I do believe the test should be approached by all parties with the utmost caution and that it should not be undertaken without safeguards. I am simply a proponent of what one genetic counselor calls "devil's advocate counseling." The Huntington's Disease Society of America has incorporated this approach into its own predictive-testing guidelines. Prospective participants are encouraged to attend numerous counseling sessions to explore fully their motivation for taking the test and their likely reactions to any possible outcome. They are urged to bring a friend with them when the results are reported. They are then encouraged to return for follow-up counseling sessions. Some consider this approach to be overkill. As genetic testing becomes more and more widespread, the counseling component will become increasingly difficult to enforce. And once the gene for Huntington's disease is isolated, the process will probably be reduced to a simple blood test. But for now, as we continue to learn about the real-life consequences of revealing potentially explosive genetic information, these guidelines or safeguards are probably necessary, not only for persons at risk who wish to be tested, but also for those doing the testing.

Yet, however carefully we tread, we will always be picking our way through a minefield. It is impossible to know what effect the predictive test for Huntington's disease will have on anyone—the affected parent, who passed on the gene, the children, who might inherit the gene, other relatives, employers, or friends. It is encouraging, but not surprising, that the Wiggins study found that just getting an answer—any answer—increased one's sense of well-being by removing the uncertainty. We are all different, but for most of us an answer helps by eliminating the daily worrying and by allowing time for planning. It is by working together that we can provide more answers to those who want them. I urge more open communication among medical professionals, families—who have a great deal to offer in the way of experience—and volunteer organizations, which can serve as a bridge between the other two groups.

Note

1. Wiggins S, Whyte P, Huggins M et al. The psychological consequences of predictive testing for Huntington's disease. *N Eng J Med* 1992: 327:1401–5.

Source: Catherine V. Hayes, "Genetic Testing for Huntington's Disease—A Family Issue," *New England Journal of Medicine* 327 (November 12, 1992): 1449–1451.

DOCUMENT 105: Gina Kolata, "Nightmare or the Dream of a New Era in Genetics?" (1993)

There are many humorous situations as well as tragedies that accompany the merger of families in marriage. When marriage occurs within a religious group that also has genetic ties, the situation can become tragic. Dor Yeshorim is a genetic screening program offered to Orthodox Jews to help remove recessive genetic diseases from the community. After voluntary screening, each person receives an identification number. When individuals begin dating or seek to marry, they can call the office to learn if they are at risk for particular diseases. This article focuses on both individual and social implications of genetic information and how such information is related to marital and reproductive decisions individuals make in light of this information.

* * *

In an ambitious attempt to eliminate common recessive diseases from their community a group of Orthodox Jews in New York and Israel is using the most advanced molecular technology to screen young people considering marriage.

It is a project that elicits strong emotions from geneticists and ethicists. Some say it is the fruit of a new genetic era. Others say it verges on a nightmare.

The leaders of the program called Dor Yeshorim, Hebrew for the generation of the righteous, say that it can serve as a model for the nation.

It does not rely on prenatal testing because abortions are generally un-acceptable to Orthodox Jews. Nor does it advise couples to avoid having children if they are at risk of passing on inherited disorders.

Campaign at High Schools

Large families of as many as 12 children are greatly desired in this community. Instead the goal is to discourage marriage or even dating between people who are at risk of having a child with a genetic disease.

Every year Dor Yeshorim representatives go to the private high schools where many Orthodox families send their children and explain to the teenagers that they can have a simple blood test to see if they carry genes for any of three diseases: Tay-Sachs, cystic fibrosis and Gaucher's disease. Then they are given a six-digit identification number. If a boy and girl want to date or if they have already started dating they are encouraged to call the New York Dor Yeshorim Central Office Hotline with their identification numbers.

Then they are told either that the match is compatible—that they are not at risk of having children with the diseases in question—or that they each carry a recessive gene that could result in a child with one of the diseases. Those couples are invited to come in for genetic counseling.

8000 Are Tested

The project is run out of a six-person office in Brooklyn and a three-person office in Jerusalem where directors of Dor Yeshorim arrange for the genetic tests at five centers in the United States and one in Israel. The tests cost $25 each and are subsidized by funds from the Department of Health and Human Services and the New York State Legislature as well as contributions from private donors.

Rabbi Josef Ekstein, a Hasid who has directed the project since its start in 1983, said that 8000 young people were tested last year for the recessive genes. So far at least 67 couples who were considering marriage have decided against it after being advised of their risk.

When Dor Yeshorim began a decade ago, it tested for just one disease, Tay-Sachs, a degenerative neurological condition that is fatal in early childhood. A few months ago, the group began testing for cystic fibrosis, which is caused by mutations in a gene that can be difficult to screen for in the general population but not in the more homogeneous population of Orthodox Jews, geneticists say. Dor Yeshorim also added a test for Gaucher's disease, a lipid-storage disease, and expects to begin screening for Canavan disease, a neurodegenerative disease. Shmuel Letkowitz, a development consultant for Dor Yeshorim, said the group wants to add tests for "anything available."

Some see this expansion as problematic. "As you move further and further away from an untreatable disease in which no one survives to cystic fibrosis or Gaucher's disease, I find the application much more troubling and much less acceptable," said Dr. Mark Siegler, an ethicist and internist at the University of Chicago School of Medicine. "It runs the risk of becoming the nightmare of the Human Genome Project," he added referring to the Federal effort to map every one of the 100,000 human genes.

Dr. Francis S. Collins, director of the Center for Human Genome Research at the National Institutes of Health in Bethesda, MD, said that parts of the program "sound just fine." For example, the genome project is expected to lead to the development of tests that would allow individuals to find out what disease-causing recessive genes they carry.

But, Dr. Collins added, other aspects of the project will "make most geneticists' hair stand on end." When there is strong pressure within a community for members to have genetic tests and to check on the genetic profiles of whomever they date, all individuals within that community may feel that they must be tested whether they want to or not. "That takes away the sacred principle of autonomy," Dr. Collins said. And as more and more genes are added to the list, some people will run the risk of being genetic wallflowers, rejected by every suitor because of the recessive genes they carry. "This is a miniature but significant version of Big Brother," Dr. Collins said. "This is a moderate nightmare."

Opponents also point out that genetic testing is full of complexities and pitfalls. Not only are there laboratory and human errors, but there are often enormous uncertainties about when and how a genetic disease might manifest itself. Many, if not most, diseases have a range of outcomes, from essentially no effect to devastating illness and, sometimes, death. Should you decide not to marry someone when the genetic disease in question might be so mild that it would never be noticed?

"We are all going to be faced with the responsibility of having this information about ourselves and how we want to have this handled in our society," said Frances Berkwits, a genetics counselor for Dor Yeshorim and the Tay-Sachs Prevention Program at Kingsbrook Medical Center in Brooklyn.

Rabbi Ekstein began Dor Yeshorim a decade ago when knowledge about genes was not as sophisticated as it is today. He and his wife saw 4 of their 10 children die of Tay-Sachs disease, watching helplessly as baby after baby developed normally for four or five months before beginning to weaken, have seizures and lose muscle control. The child would become blind and paralyzed, and after a few years would die.

At first, the rabbi said, he never spoke about his Tay-Sachs babies. "When something like this happens we try to cover it up," he said. Like many families in his religious community, he said, he and his wife were afraid that if anyone knew they had a sick child, no one would want to marry the healthy ones.

About a year and a half after his fourth child died, Rabbi Ekstein said he realized that the reason God gave him four children with Tay-Sachs disease was so that he could help others prevent the disease in their families. He began Dor Yeshorim.

At first, the program met with stubborn resistance. Mr. Lefkowitz said that the general attitude was, "If God wants me to have a Tay-Sachs child, I'll have a Tay-Sachs child."

But Rabbi Ekstein and his supporters went from rabbi to rabbi, convincing them to encourage people to be tested. The Dor Yeshorim proponents argued that when a child wants to marry, the parents make sure that he or she will be marrying into a religious family with appropriate status in the community and with some money. "That you don't leave to God," Mr. Lefkowitz said. "So why leave this to God? God has enough to do."

The first year, 45 people agreed to be tested. The next year, 250 were tested. "Then it just mushroomed," Mr. Lefkowitz said.

Mr. Lefkowitz's daughter, Mindy Erlanger, was among those tested. Mrs. Erlanger, now a 21-year-old bookkeeper in Brooklyn, said that when she was 16, she had the Tay-Sachs test, the only disease Dor Yeshorim was screening for at that time. Two years later, when she was dating a young man she felt she might consider marrying, they called the registry to see if they were carriers of the gene. They were not and the following year they were married. They now have a healthy son. "All my friends went through the same routine," Mrs. Erlanger said. But, she added, if it had turned out she and her prospective husband were both carriers of the Tay-Sachs gene, she would not have married him.

A man who asked not to be identified broke off his engagement when he and his fiancee learned they each carried the Tay-Sachs gene. He soon began seeing another young woman and after their third date the two called for her test results. She did not have the gene, which meant that their children would not have the disease. The two were married and now have three healthy children. "Coming home, having a kid run into my arms, having three healthy children is a joy," the man said. He added that the Dor Yeshorim program "literally saved my life."

Another young woman stopped dating two men after discovering that they carried Gaucher's disease, even though her parents said that she should not make her decision on that basis alone. "My parents are like, 'If you both love each other, go for it,'" she said. "I wasn't willing to take that gamble." Dr. Michael Kaback, a medical geneticist at the University of California at San Diego who directs the Tay-Sachs program for California, said he has no objections to the Tay-Sachs program if the testing is accurate and the people are fully informed. But, he adds, Tay-Sachs is an exception because, unlike the vast majority of genetic diseases, its consequences are almost always severe and tragic. "When they start packaging other things in there, I get real concerned," Dr. Kaback said. For example, Dr. Kaback questions the screening for Gaucher's disease, which is characterized by an enzyme deficiency that can make the liver and spleen enlarge, the bones become brittle and the bone marrow fail to function normally, leading to severe anemia. The first symptom usually does not surface until about age 45. Many people never know they have the disease, and it may be identified only on autopsy, when a doctor notices that the spleen is enlarged. A minority of people have severe and painful symptoms appearing in childhood.

The disease can be treated with the drug Ceredasee. The drug now costs $100,000 a year, but Dr. Kaback said that price is widely expected to drop as new methods of production become available. So, he asked, does it make sense for a man and woman not to marry because they both carry a Gaucher's gene? And how are they being counseled?

As more genes are found, the problems multiply. "What are we going to do when we have 50 or 100 genes?" Dr. Kaback said. In fact, he added, if Dor Yeshorim waits long enough there will be so many recessive disease-causing genes identified that "every single marriage will be prevented."

Dr. Kaback argues that scientists have oversimplified the benefits of a genetic screening. "This mentality, unfortunately, has been fostered in some degree, by the scientific community," he said, explaining that the message has been, "If a test exists, you should use it."

But Mr. Ekstu said that the uncertainties in the disease outcomes make it even more desirable to test people before they marry. "With Tay-Sachs, there may be an ethical reason to abort," he said. "But there is no ethical reason to abort a Gaucher's baby." He believes, however, that couples should be able to decide before they marry whether they want to risk having a baby with the disease. "Everyone has to make that decision—do I want to take that risk or not?" he said.

Source: Gina Kolata, "Nightmare or the Dream of a New Era in Genetics?" *New York Times*, December 7, 1993, A1 and C3.

DOCUMENT 106: Gina Kolata, "Advent of Testing for Breast Cancer Genes Leads to Fears of Disclosure and Discrimination" (1997)

The difficulties created by the use of genetic screening are shown in the case reported in this document. A woman was told she had a mutation that could cause breast cancer. The woman opted for a double mastectomy as a preventive measure. Prior to the operation she submitted a claim without revealing she had the test but reporting a family history of breast cancer; the claim was rejected because the company did not pay for preventive medicine. She then had the physician submit the genetic test results, only to be told she would not qualify for payment for surgery because of the existence of a preexisting medical condition. This case, and the discussion it raised, points to the many dilemmas we face both in the practice of medicine and in obtaining health insurance and securing payment for various treatments. The more that is known about us, the less qualified for insurance we may become. (See Part V.)

* * *

Barbara Weber is still reeling from the experience of a patient who told her insurance company that she had a mutated gene that could cause breast cancer.

The woman had been tested for the mutation at Dr. Weber's clinic at the University of Pennsylvania, and Dr. Weber told her the results. Since some studies indicate that women with the mutated gene have a 90 percent chance of developing breast cancer, the woman wanted both of her breasts removed right away. Before she had the operation, she submitted a claim to her insurance company, Dr. Weber said, not disclosing that she had had the genetic test but reporting a strong family history of breast cancer.

The company turned her down, Dr. Weber said, on the ground that it did not pay for preventive medicine. So, at the woman's request, Dr. Weber submitted the woman's genetic test results. At that point, Dr. Weber said, the company told the woman that it would not pay for the surgery because she had a preexisting condition—a genetic defect— when she took out her health policy.

"It was absolutely unbelievable," Dr. Weber said. She said that the woman, who did not return telephone calls requesting an interview, had the surgery anyway. Afterward, Dr. Weber said, when pathologists examined the woman's breast tissue, they found a cancerous tumor that had been missed by mammograms.

It is cases like this and fears of similar treatment that are convincing some women and researchers that it might be too dangerous to put genetic testing results on medical charts and in clinical records, where privacy cannot be insured. Women worry that insurers will raise their rates, or refuse to insure them, that employers will not hire them or promote them, and even that friends and family members might treat them differently if they knew that they were tainted with a deadly gene.

Fears of genetic testing have been expressed by some people for several years, starting with tests for rare genetic diseases like Huntington's.

But the recent discovery of two genes that can cause breast and ovarian cancer when mutated has made the issue more pressing.

Not only is breast cancer the disease that women fear the most, but the mutations in genes known as BRCA1 and BRCA2 are relatively common in Ashkenazi Jewish populations (of Eastern European descent), affecting as many as 1 in 50 women.

In addition, many women who are not of Jewish descent but have a strong family history of breast or ovarian cancer also inherited mutated genes. In the last year, several commercial laboratories have begun offering tests for breast cancer gene mutations and major medical centers across the country have set up their own testing programs.

Some legal experts say the fears of discrimination may be exaggerated. Federal legislation that will go into effect on July 1 addresses part of the issue, focusing on people who are covered under group medical plans. The law states that if an individual was in such a plan for at least 12 months and had a genetic condition diagnosed in the past 6 months, a new insurer cannot use that genetic information to deny or limit coverage.

Alexander Morgan Capron, an ethicist and law professor at the University of Southern California, said insurers could charge people with genetic conditions more for insurance. But, Mr. Capron added, 20 states, including New Jersey, have recently passed laws preventing health maintenance organizations and health insurance companies from charging people more because they have a gene mutation. In addition, the Americans with Disabilities Act prevents employers from discriminating against people who have disease-causing genetic mutations. Employers who insure their employees are exempt from the state laws, Mr. Capron said.

Richard Coorsh, a spokesman for the Health Insurance Association of America, said that people's fears that insurance companies would discriminate against them might be overblown. Many people are insured through their employers as part of large groups, and they would be covered even if they did have a cancer-causing gene.

But Mr. Coorsh said that those who change jobs or lose their jobs and have to seek insurance as individuals "will probably end up having to pay a lot more money" for health insurance if they test positive for a cancer gene. There have been few systematic studies to assess how real the risk of genetic discrimination is, but one survey, published in October in the journal *Science*, showed that the fear, at least, is widespread. The study, by Dr. E. Virginia Lapham of the Georgetown University School of Medicine and her colleagues, involved 332 people who belonged to support groups for families with a variety of genetic disorders. Of them, 25 percent believed they were denied life insurance because of their disorder, 22 percent thought they were denied health insurance, and 13 percent believed they were not hired for a job or that they lost a job because of the disorder.

Whether or not these suspicions are justified, they are having an effect. Some researchers at major medical centers are worried enough to work under a cloak of secrecy. The test for the breast cancer genes is under the aegis of a research program which allows doctors to keep the results secret, coded and encased in locked files.

Some researchers advise women who are tested not to tell even their private doctor of their test results if the doctor insists on putting all relevant information in a woman's medical record. Some doctors agree not to write down test results, relying instead on their memories.

At treatment centers, which, unlike research centers, are not permitted to hide test results, some women have used aliases to protect their privacy. "If we ever needed proof that the system is broken, this is it," said Dr. Francis Collins, director of the National Center for Human Genome Research in Bethesda, MD. "The system forces people to take drastic steps to protect themselves. It is putting a terrible burden on patients."

Doctors, too, are placed in an untenable situation, Dr. Collins said. "You are forced sometimes to have interactions with insurance agents or with other physicians or with H.M.O.'s where you would have to pretend you don't have the information," Dr. Collins said. "It's a very strange dilemma—to choose between patient confidentiality and telling the truth."

It is, said Dr. Thomas Murray, the director of the Center for Biomedical Ethics at Case Western Reserve University in Cleveland, "a classic ethical quandary, and one with no obvious solution."

Moreover, some cancer researchers note, there is a legitimate reason why insurers and employers might want to avoid people with genes that can cause cancers: the insurers because of the extra cost of treating them and the employers because of the extra insurance cost and of lost productivity. "I'm not sure you can totally blame insurance companies or employers," said Dr. C. Kent Osborne, a cancer researcher at the University of Texas in San Antonio. "It's a bigger problem than that—it's something we have to deal with as a society."

In the meantime, doctors and patients are trying to act expediently.

Dr. Funmi Olopade, who directs the breast cancer genetic testing program at the University of Chicago, was interviewed just after an appointment with a 25-year-old woman, who came to learn her test results. The woman was found to have a genetic mutation that can cause breast and ovarian cancer. The university keeps these records secret. But, Dr. Olopade said, the woman asked her if she could tell her private doctor of her test and what it disclosed. Dr. Olopade suggested that the young woman start by asking her doctor if he felt he had to put everything he knew about a patient into the woman's medical records. If so, Dr. Olopade said, the woman might not want to tell.

And with good reason, said Dr. Mark Siegler, who directs the ethics program at the University of Chicago. More than a decade ago, before the proliferation of electronic records, before the merging of health maintenance organizations that both pay for and care for patients, he investigated the complaint of a woman who had come to the hospital for elective gall bladder surgery and had said that too many people were looking at her medical records.

"I started to ask how many people had legitimate access to her chart," Dr. Siegler said. "I stopped counting at 75." He concluded, he said, "that confidentiality really didn't exist." "That is even more true today," he added.

Some doctors, like Dr. John Glick, the director of the University of Pennsylvania's cancer center, say they will not put genetic testing information in a woman's medical records if she asks that it be kept out. Instead, Dr. Glick said, he simply remembers what the woman said about her test results.

Other doctors refuse to go along with such requests. One woman, who tested positive for the breast cancer gene at the University of Pennsylvania, said her private doctor told her it would be dishonest not to include the test result in her medical records. So, the woman said, "I asked that my medical records be closed and I changed doctors." The woman said she was worried about health insurance. "The company my husband works for just got bought out, so things aren't as steady as they used to be," she said. "You can survive breast cancer, but you can't survive if you don't have insurance."

Dr. Henry Lynch, who directs the Cancer Center at Creighton University in Omaha, tests women for the breast cancer gene as part of a research program in which he can keep the results secret, and as part of a clinical program, where he cannot. But some women in his clinical program use aliases. Others pay for tests out of their own pockets, hoping to keep the information out of the hands of their insurance companies.

Dr. Weber said that she was confronting a new ramification of women's fears of discrimination. Women who were tested as part of her research program are refusing to participate in long-term studies that would assess their risk of developing cancer and the success of interventions, like frequent mammograms, or hormone treatments, or surgery to remove their breasts or ovaries. The women, Dr. Weber said, fear that it would be a red flag on their medical records to participate in the study. Their mammogram results and any biopsies, for example, would have to be sent to the university's research program.

The only possibility of learning if there is a way to prevent cancer in women who are genetically predisposed, Dr. Weber said, is to do long-term studies. If no one will participate, she said, what is the point in testing women to see if they have the cancer genes?

Source: Gina Kolata, "Advent of Testing for Breast Cancer Genes Leads to Fears of Disclosure and Discrimination," *New York Times*, February 4, 1997, C1 and C3.

DOCUMENT 107: Kathy L. Hudson, Karen H. Rothenberg, Lori B. Andrews, Mary Jo Ellis Kahn, and Francis S. Collins, "Genetic Discrimination and Health Insurance: An Urgent Need for Reform" (1995)

Most individuals who have health insurance in this country obtain it through private insurance companies. Typically insurance companies have used physical exams as a way of either denying insurance to individuals at high risk or insuring at-risk clients only with extremely high premiums. Developments in genetics, such as those described in the previous document, have raised the stakes in this practice because we can now more precisely identify individuals who have a predisposition to various diseases as well as those with specific genetic diseases or with late-onset diseases. Researchers and the general public are beginning to question whether the exclusion of such individuals from health insurance is discriminatory. The authors of this document make several recommendations about the provision of health insurance to such populations.

* * *

The accelerated pace of gene discovery and molecular medicine portend a future in which information about a plethora of disease genes can be readily obtained. As at-risk populations are identified, research can be done to determine effective prevention and treatment strategies that will lower the personal, social, and perhaps the financial costs of disease in the future. We all carry genes that predispose to common illnesses. In many circumstances knowing this information can be beneficial, as it allows individualized strategies to be designed to reduce the risk of illness. But, as knowledge about the genetic basis of common disorders grows, so does the potential for discrimination in health insurance coverage for an ever increasing number of Americans.

The use of genetic information to exclude high-risk people from health care by denying coverage or charging prohibitive rates will limit or nullify the anticipated benefits of genetic research. In addition to the real and potentially devastating consequences of being denied health insurance, the fear of discrimination has other undesirable effects. People may be unwilling to participate in research and to share information about their genetic status with their health care providers or family members because of concern about misuse of this information. As genetic research progresses, and preventive and treatment strategies are developed, it will be increasingly important that discrimination and the fear of discrimination not be a roadblock to reaping the benefits. To address these issues, the National Institutes of Health–Department of Energy (NIH-DOE) Working Group on Ethical, Legal, and Social Implications (ELSI) of the Human Genome Project and the National Action Plan on Breast Cancer have jointly developed a series of recommendations for state and federal policy makers which are presented below.

In the past, genetic information has been used by insurers to discriminate against people. In the early 1970s, some insurance companies denied coverage and charged higher rates to African Americans who were carriers of the gene for sickle cell anemia (1). Contemporary studies have documented cases of genetic discrimination against people who are healthy themselves but who have a gene that predisposes them or their children to a later illness such as Huntington's disease (2). In a recent survey of people with a known genetic condition in the family, 2296 indicated that they had been refused health insurance coverage because of their genetic status, whether they were sick or not (3).

As a case example, Paul (not his real name) is a healthy, active 4-year-old, but he has been twice denied health insurance. Paul's mother died in her sleep of sudden cardiac arrest when Paul was only 5 months old. Paul's maternal uncle also died of sudden cardiac arrest when he was in his twenties. After these sudden and unexpected deaths, Paul's family began a hunt to discover the cause. Their search finally led to a research geneticist who was able to determine that several family members, including Paul and his mother, carried an alteration in a gene on chromosome 7. This gene is one of several genes that causes the long QT syndrome, so-called because of the distinctive diagnostic pattern on an electrocardiogram.

Several years ago, Paul's father, Bob, lost his job and with it the group policy that provided health insurance coverage for Paul and him. Paul's father has repeatedly applied for a family health insurance policy with a major insurance company. The company agreed to cover Bob but refused to issue a family policy that would cover Paul because he has inherited the altered gene for the long QT syndrome from his mother.

The story of Jackie and Emma further illustrates the social, ethical, and legal dilemmas presented by the revelation of genetic information. Sisters Jackie and Emma, along with many other members of their family, have been tested as part of a research protocol for alterations in the gene, BRCA1, that confers hereditary susceptibility to breast and ovarian cancer. Both were offered an opportunity to learn the results of their genetic tests and both accepted. They each learned they carry an altered form of the gene, putting them at increased risk for breast and ovarian cancer.

After finding out the results of her genetic test, Emma had a mammogram that showed a very small lesion in her breast. A subsequent biopsy revealed carcinoma and Emma decided to proceed with a bilateral mastectomy because of the substantial risk of cancer arising in the opposite breast. Her lymph nodes were negative for cancer, so her prognosis for cure is very good. Sister Jackie also tested positive for the same alteration in the BRCA1 gene though no cancer was detected. Although the benefit of prophylactic mastectomy in reducing the risk for breast cancer is not yet known, she decided to have a bilateral prophylactic mastectomy. Emma and Jackie feel strongly that they have benefited from knowing this genetic information but are fearful that it will be used against them and their family by insurers and employers. They both keep their genetic status secret and are so fearful of losing their health insurance that they used assumed names when sharing their story at a recent workshop on genetic discrimination (4).

Emma and Jackie's story is not unique. An estimated 1 in 500 women carry a mutation in the BRCA1 gene that may confer as much as an 85% chance of breast cancer and a 50% chance of ovarian cancer (5). Although substantial uncertainty exists about the relative value of the available options (surgery compared with intensive surveillance) for a woman with a BRCA1 mutation, it is likely that ultimately this information will be medically useful.

Health Insurance in the United States

Because of high costs, insurance is essentially required to have access to health care in the United States. Over 40 million people in the United States are uninsured (6). Group insurance, individual insurance, self-insurance, and publicly financed insurance (for example, Medicare and Medicaid) are the principal forms of health insurance in the United States for the ±240 million Americans with coverage. Most people get their health insurance through their employer. Many employers provide health insurance coverage through self-funded plans in which the employer, either directly or through a third party, provides health insurance coverage. For individuals and small groups, insurance providers use medical history as well as individual risk factors, such as smoking, to determine whether to provide coverage and under what terms. This is known as underwriting. Insurers argue that underwriting is essential in a voluntary market to prevent "adverse selection," in which individuals elect not to purchase insurance until they are already ill or anticipate a future need for health care. Insurers fear that individuals will remain uninsured until, for example, they receive a genetic test result indicating a predisposition to some disease such as breast or colon cancer.

In the absence of the ability to detect hereditary susceptibility to disease, the costs of medical treatment have been absorbed under the current health insurance system of shared risk and shared costs. Today, our understanding of the relation between a mispelling in a gene and future health is still incomplete, thus limiting the ability of insurers to incorporate genetic risks into actuarial calculations on a large scale. As genetic research enhances the ability to predict individuals' future risk of diseases, many Americans may become uninsurable on the basis of genetic information.

State and Federal Initiatives

A recent survey has shown that a number of states have enacted laws to protect individuals from being denied health insurance on the basis of genetic information (7). The first laws addressing genetic discrimination were quite limited in scope and focused exclusively on discrimination against people with a single genetic trait such as sickle cell trait (8). Since the Human Genome Project was launched in 1990, eight states have enacted some form of protection against genetic discrimination in health insurance. The recently enacted state laws are not limited to a specific genetic trait but apply potentially to an unlimited number of genetic conditions. These state laws prohibit insurers from denying coverage on the basis of genetic test results, and prohibit the use of this information to establish premiums, charge differential rates, or limit benefits. A few of these states, including Oregon and California, integrate protection against discrimination in insurance practices with privacy protections that prohibit insurers from requesting genetic information and from disclosing genetic information without authorization.

Two factors limit the protection against discrimination afforded by current state laws. First, the federal Employee Retirement Income Security Act exempts self-funded plans from state insurance laws. Nationwide, over one-third of the nonelderly insured population obtains health insurance coverage through a self-funded plan. Second, nearly all of the state laws focus narrowly on genetic tests, rather than more broadly on genetic information generated by family history, physical examination, or the medical record (7). Limiting the scope of protection to results of genetic tests means that insurers are only prohibited from using the results of a chemical test of DNA, or in some cases, the protein product of a gene. But insurers can use other phenotypic indicators, patterns of inheritance of genetic characteristics, or even requests for genetic testing as the basis of discrimination. Meaningful protection against genetic discrimination requires that insurers be prohibited from using all information about genes, gene products, or inherited characteristics to deny or limit health insurance coverage.

No federal laws are currently in place to prohibit genetic discrimination in health insurance (9). The Clinton Administration's proposal to reform the health care system and provide health insurance for all Americans did prohibit limiting access or coverage on the basis of "existing medical conditions or genetic predisposition to medical conditions" (10). Congressional efforts to reform the health care system in 1995 have been much more modest and are targeted at guaranteeing access, portability, and renewability of coverage and at leveling the playing field in the insurance market so that the same rules apply to insured and self-funded plans. Recent federal health insurance reform proposals attempt to guarantee the availability of health care by prohibiting insurers from denying coverage on the basis of health status, medical condition, claims experience, or medical history of a participant. Most of the proposals permit exclusions for pre-existing conditions, but these are time limited.

It is not clear if the current health insurance reform proposals would prohibit insurers from denying coverage on the basis of genetic information. Genetic information is distinct from other types of medical information because it provides information about an individual's predisposition to future disease. In addition, genetic information can provide clues to the future health risks for an individual's family members. If enacted, current health reform proposals would prohibit denying insurance to those currently suffering from disease or with a past history of disease. But these proposals may not protect people like Paul, who are healthy but have a genetic predisposition to disease, from being refused insurance coverage. Current proposals also may fail to protect couples who, although healthy themselves, carry the gene for a recessive disorder such as cystic fibrosis that might affect their children or future children.

Recommendations

Planners of the Human Genome Project recognized from the beginning that maximizing the medical benefits of genome research would require a social environment in which health care consumers were protected from discrimination and stigmatization based on their genetic make-up. Genome programs at both the DOE and the National Center for Human Genome Research, a component of NIH, have each set aside a portion of their research budget to anticipate, analyze, and address the ELSI of new advances in human genetics. The original planners also created the NIH-DOE ELSI Working Group, which has a broad and diverse membership including genome scientists; medical geneticists; experts in law, ethics, and philosophy; and consumers to explore and propose options for the development of sound professional and public policies related to human genome research and its applications. The ELSI Working Group has long been involved in discussions about the fair use of genetic information. In a 1993 report, "Genetic Information and Health Insurance" (11), the ELSI Working Group recommended a return to the risk-spreading goal of insurance. The Working Group suggested that individuals be given access to health care insurance irrespective of information, including genetic information about their past, current or future health status. Because denial of insurance coverage for a costly disease such as breast cancer may prove to be a death sentence for many women, the National Action Plan on Breast Cancer (NAPBC), a public-private partnership designed to eradicate breast cancer as a threat to the lives of American women, has identified genetic discrimination in health insurance as a high priority (12).

Building on their shared concerns, the NAPBC (13) and the ELSI Working Group (14) recently cosponsored a workshop on genetic discrimination and health insurance (4). Scientists, representatives from the insurance industry, and members of the ELSI Working Group and the NAPBC participated in the 1 day session. On the basis of the information presented at the workshop, the ELSI Working Group and the NAPBC developed the following recommendations and definitions for state and federal policymakers to protect against genetic discrimination.

1. Insurance providers should be prohibited from using genetic information, or an individual's request for genetic services, to deny or limit any coverage or establish eligibility, continuation, enrollment, or contribution requirements.

2. Insurance providers should be prohibited from establishing differential rates or premium payments based on genetic information or an individual's request for genetic services.

3. Insurance providers should be prohibited from requesting or requiring collection or disclosure of genetic information.

4. Insurance providers and other holders of genetic information should be prohibited from releasing genetic information without prior written authorization of the individual. Written authorization should be required for each disclosure and include to whom the disclosure would be made.

The definitions are as follows. Genetic information is information about genes, gene products, or inherited characteristics that may derive from the individual or a family member. Insurance provider means an insurance company, employer, or any other entity providing a plan of health insurance or health benefits including group and individual health plans whether fully insured or self-funded.

These recommendations have been endorsed by the National Advisory Council for Human Genome Research (NACHGR) (15). The NACHGR stresses the positive value of genetic information for improving the medical care of individual patients and the need to ensure the freedom of patients and their health care providers to use genetic information for patient care. The NACHGR views the elimination of the use of genetic information to disseminate against individuals in their access to health insurance as a critical step toward these goals.

The ability to obtain sensitive genetic information about individuals, families, and even populations raises profound and troubling questions about who will have access to this information and how it will be used. The recommendations presented here for state and federal policymakers are intended to help ensure that our current social, economic, and health care policies keep pace with both the opportunities and challenges that the new genetics present for understanding the causes of disease and developing new treatment and preventive strategies.

REFERENCES AND NOTES

1. L. Andrews, *Medical Genetics: A Legal Frontier* (American Bar Foundation, Chicago, IL, 1987).

2. P. R. Billings et al., *Am. J. Hum. Gen.* 50, 476 (1992).

3. E. V. Lapham (Georgetown University) and J. O. Weiss, *The Alliance of Genetic Support Groups*. Human Genome Model Project, preliminary results of a survey of persons with a genetic disorder in the family.

4. "Genetic discrimination and health insurance: A case study on breast cancer." Bethesda, MD, 11 July 1995, workshop sponsored by the National Action Plan on Breast Cancer (NAPBC) and the NIH-DOE Working Group on the Ethical, Legal, and Social Implications of Human Genome Research.

5. D. F. Easton et al., *Am. J. Hum. Genet.* 52, 678 (1993); D. Ford et al., *Lancet* 343, 692 (1994).

6. *Employee Benefit Research Institute Special Report SR-28*, issue brief number 158, February 1995.

7. K. H. Rothenberg, *J. Law Med. Ethics*, in press.

8. North Carolina, NC ST: 58–65–70 (1975); Florida, FL ST: 626.9707 (1978); Alabama, AL ST: 27–5–13 (1982). In 1987, Maryland passed a law, Art. 48A, 223(b)(4), prohibiting health insurers from discrimination in rates based on genetic traits unless there was "actual justification."

9. In March 1995, the U.S. Equal Employment Opportunity Commission (EEOC) released official guidance on the definition of the term "disability." The EEOC's guidance clarifies that protection under the Americans with Disabilities Act (ADA) extends to individuals who are discriminated against in employment decisions solely on the basis of genetic information about an individual. For example, an employer who makes an adverse employment decision on the basis of an individual's genetic predisposition to disease, whether because of concerns about insurance costs, productivity, or attendance, is in violation of the ADA because that employer is regarding the individual as disabled. Issuance of the

EEOC's guidance is precedent setting; it is the first broad federal protection against the unfair use of genetic information.

10. *Health Security Act*, Section 1516, S. 1755/HR 3600.

11. "Genetic information and health insurance. Report of the task force on genetic information and insurance" (NIH-DOE Working Group on the Ethical, Legal, and Social Implications of Human Genome Research, 10 May 1993).

12. The NAPBC has as its mission to reduce the morbidity and mortality from breast cancer and to prevent the disease. Specific goals include the following: (i) to promote a national effort to establish and address priority issues related to breast cancer etiology, early detection, treatment, and prevention; (ii) to promote and foster communication, collaboration, and cooperation among diverse public and private partners; and (iii) to develop strategies, actions, and policies to improve breast cancer awareness, services, and research.

13. NAPBC steering committee: Susan J. Blumenthal (co-chair), Zora Kramer Brown, Doris Browne, Anna K. Chacko, Francis S. Collins, Nancy W. Connell, Kay Dickersin, Arlyne Draper, Nancy Evans, Harmon Eyre, Leslie Ford, Janyce N. Hedetniemi, Mary Jo Ellis Kahn, Amy S. Langer, Susan M. Love, Alan Rabson, Jane Reese-Coulbourne, Irene M. Rich, Barbara K. Rimer, Susan Sieber, Edward Sondik, and Frances M. Visco (co-chair). NAPBC hereditary susceptibility working group: Kathleen A. Calzone, Francis S. Collins (co-chair), Sherman Elias, Linda Finney, Judy E. Garber, Ruthann M. Giusti, Jay R. Harris, Joseph K. Hurd Jr., Mary Jo Ellis Kahn (co-chair), Mary-Claire King, Caryn Lerman, Mary Jane Massie, Paul G. McOonough, Patricia O. Murphy, Philip D. Noguchi, Barbara K. Rimer, Karen H. Rothenberg, Karen K. Steinberg, and Jill Stopfer.

14. ELSI working group: Betsy Anderson, Lori Andrews (chair), James Bowman (dissenting), David Cox, Troy Duster (vice chair), Rebecca Eisenberg, Beth Fine, Neil Holtzman, Philip Kitcher, Joseph McInemey, Jeffrey Murray, Dorothy Nelkin, Rayna Rrapp, Marsha Saxton, and Nancy Wexler.

15. NACHGR council members: Anita Allen, Lennette J. Benjamin, David Botstein, R. Daniel Camerini-Otero (dissents with recommendation 3), Ellen W. Clayton, Troy Duster, Leroy E. Hood, David E. Housman, Richard M. Myers, Rodney Rothstein, Diane C. Smith, Lloyd M. Smith, M. Anne Spence, Shirley M. Tigham, and David Valle.

Source: Kathy L. Hudson, Karen H. Rothenberg, Lori B. Andrews, Mary Jo Ellis Kahn, and Francis S. Collins, "Genetic Discrimination and Health Insurance: An Urgent Need for Reform," *Science* 270 (October 20, 1995): 391–393.

DOCUMENT 108: Jeffrey M. Leiden, "Gene Therapy— Promise, Pitfalls and Prognosis" (1995)

Genetic diagnosis and gene therapy are still an evolving science, as noted in the following document. The author here makes recommendations for future research in this area, providing a look at what the future may hold.

* * *

Any successful form of gene therapy must combine an appropriate disease target with a gene-delivery system that programs temporal and spatial patterns of recombinant-gene expression that produce long-term therapeutic results and little or no toxicity. Many of the early clinical trials targeted recessive disorders involving a single gene. The reasons for this are straightforward. The genes responsible for many of these diseases have been cloned, there is a dearth of viable therapeutic options for patients with these devastating disorders, and it is clear that the long-term expression of a normal copy of the relevant gene would correct the abnormality. However, these diseases are particularly difficult to treat with somatic gene therapy. Successful gene therapy for both Duchenne's muscular dystrophy and cystic fibrosis requires the delivery and long-term expression of the appropriate gene to large numbers of cells throughout inflamed and fibrotic target tissues.

Given these difficulties and the recent progress in developing new systems of in vivo gene delivery, it may be worthwhile to reconsider the initial targets of gene therapy. For example, the finding that intramuscular injection of plasmid DNA can program recombinant gene expression in skeletal myocytes (1) has led to the development (in animals) of novel DNA-based vaccines for infectious diseases. Such an approach should be particularly easy to translate into human therapy, because it requires only local, transient, and low-level expression of recombinant genes in relatively few cells. Similarly, the observation that the local delivery by a catheter of retinoblastoma or herpes simplex virus thymidline kinase genes can be used to reduce the rate of restenosis after balloon angioplasty holds promise for the treatment of coronary-artery restenosis.

What is the future of gene therapy for diseases such as cystic fibrosis, Duchenne's muscular dystrophy, and inherited serum protein deficiencies? Our preclinical and clinical experience demonstrates the need for new or improved gene-delivery systems to treat such disorders. Promising new approaches include methods of modifying adenovirus vectors, as well as immunosuppressive regimens that prolong recombinant-gene expression in animals. (2, 3) Nevertheless, it is important to remember that gene therapy is truly in its infancy and that the current tools are quite crude. Thus, it is essential that we continue generous funding of research initiatives designed to study issues of basic scientific importance to the field. Unexpected findings from such basic research efforts often lead to the development of new and improved approaches to gene therapy.

Early clinical experience with gene therapy reminds us of important ethical issues. Peer review is essential to evaluate the scientific and ethical basis of gene-therapy trials. It is equally important that we (and the lay press) present and publish the results of such trials in a balanced way and temper our enthusiasm with practical reality. Despite some pitfalls, the promise of gene therapy is intact. There is good reason to be optimistic about the ultimate success of this treatment.

References

1. Wolff J A, Malone R W, Williams P et al. Direct gene transfer into mouse muscle in vivo. *Science* 1990:247:1465–8.

2. Verma I M. Gene therapy: hopes, hypes, and hurdles. *Mol Med* 1994: 1:2–3.

3. Dai Y, Schwarz E M, Gu D, Zhang W W, Sanetnick N, Verma I M. Cellular and humoral immune responses to adenoviral vectors containing factor IX gene: tolerization of factor IX and vector antigens allows for long-term expression. *Proc Natl Acad Sci U S A* 1995:92:1401–05.

Source: Jeffrey M. Leiden, "Gene Therapy—Promise, Pitfalls and Prognosis," *New England Journal of Medicine* 333 (September 28, 1995): 871–872.

Part VII

Ethical Issues in
Genetic Engineering

Along with all the excitement and promise of the developments in genetics come a set of ethical problems that are both personal and social. Many of these problems have been alluded to in the documents in the previous sections. Here, the documents will focus specifically on these ethical issues and provide a variety of analyses of them.

DOCUMENT 109: Willard Gaylin, "The Frankenstein Factor" (1977)

Public misperception of scientific research, particularly genetic research, is a problem. Two main points of this problem are examined by the author of the following document, which was written in the early days of genetic engineering. First, high-tech interventions are more feared than low-tech interventions—even though the less-feared low-tech interventions may have the same consequences as the more-feared high-tech interventions. Second, research perceived as changing human nature will be perceived as more fearful than other equally risky and dangerous research. The author of the following document argues that the conscious or unconscious appearance of these elements, which he calls the Frankenstein factor, has the capacity to misinform the public and misrepresent scientific research.

* * *

Recombinant DNA is becoming the Farrah Fawcett-Majors [the Heather Locklear of the 1970s] of scientific issues of the day. It is impossible to get through a week's reading of newspapers and magazines (let alone scientific publications) without encountering yet another insider's view of this glamorous subject. If recombinant DNA research is destroyed by the publicity surrounding it, it will be due to anxiety, not boredom. The fate of such extraordinarily promising research ought not be influenced by emotional, rhetorical or irrational reasons.

It seems clear to me that there is an unanalyzed element coloring the debate about this research. Within my limits as a man unsophisticated in molecular biology, I have attempted to read everything I could understand about the process. That its potential is extraordinary and that it represents a major new direction of research seems unquestionable. There are risks, as always when one is entering uncharted areas. My own judgment—and I offer my conclusions with no great assurance or authority—is that the predictable benefits are worth the predictable risks. I recognize that men of goodwill disagree, and that the "experts" are divided, thus further confusing the public. But I suspect that some of the opposition to recombinant DNA emerges from factors beyond the specific weighing of risks and benefits that one would encounter in other forms of research. That extra, additional and unanalyzed variable, which I believe to be distorting the public debate, I will call the "Frankenstein Factor."

The Frankenstein Factor consists of two major components, which together inevitably enhance the normal anxiety about risks in a given research. The first, high-technology research with human beings, with
its element of mystery for the layman, will be feared more than low-
technology technics, which effect the same ends. For example, experiments to modify behavior with electrodes will be feared more than
experiments with drugs, and the latter more than experiments with op-
erant conditioning—even though all three bypass rationality and choice
in an effort to control human actions.

Secondly, research that is seen as changing or controlling the "nature"
of the species or controlling behavior will inevitably be received with
more fear than other equally risky research. In some procedures the human being is at risk as a test medium only—e.g., in testing side effects
of drugs. Even this form of research will be seen as less threatening—
even if more dangerous—than procedures in which the change in the
person is the end goal, not just the side effect.

Genetic engineering represents such an area. At this point it might be
wise to compare the debate over recombinant DNA with another issue
that captured headlines, and the public imagination, some few years
back: psychosurgery. Here, too, one sensed an emotional overlay to
much of the discussion. There was, of course, legitimate concern about
abuse of psychosurgery (as, indeed, there are legitimate concerns about
recombinant DNA). It may even be that those opposed to both procedures may be correct. What is disturbing in both cases is the quality of
the debate, which seems to supply an extra dimension of anxiety and
concern.

It is not simply a "failure to communicate"; rather, in these areas,
anxiety levels will make this particular research especially vulnerable to
political rhetoric. It is the unconscious aspects of the Frankenstein Factor
that will move public opinion, even though that fear will be rationalized
by overt use of more realistic arguments.

Obviously, electrode implantations or surgical ablation of brain sections as a direct means of political control are cumbersome and unlikely—much less a threat, for example, than conditioning or, for that
matter, drugs. Such an individualized and dramatic procedure as psychosurgery is hardly suited to the enslavement of masses or robotization
of opposing political leaders. Drugs, brain washing by control of television, exploitation of fears through forms of propaganda and indoctri-
nation through the sources of education—particularly of preschool
education or neonatal conditioning, as suggested by Skinner—all seem
more likely methods of totalitarian control. Why, then, the extraordinary
fuss about psychosurgery? It cannot just be explained by aversion to
surgery; compare the legislative flurry here with the almost reverential
passivity about heart transplants.

Success in behavior modification and genetic engineering research is more likely to bring dejection than in other kinds of biologic experimentation. Devices that save and extend life aggrandize both the discoverer and man in general with the suggestion that such control of death, although still not the immortality of God, is a cut above the helplessness of the general animal host. Behavior manipulation, on the other hand, reasserts man's kinship with the pigeon, the rat and the guinea pig. The more technologic the control devices, the more mechanical the method—the scarier it all seems.

Genetic engineering is even more terrifying, with the implications of tampering with the stuff of life itself and in the process reducing people to the level of manufactured items. It is the Frankenstein myth realized. When Mary Shelley first devised her story in 1818, the scientific impact on society was just at its beginning. The idea of one human being fabricating another was purely metaphorical. The process was assumed to be impossible. It was, to use her words, "a ghost story." But the inconceivable has become conceivable, and in the 20th century human beings are indeed patched together out of parts. We sew on detached arms and fix shattered hips in place with metal spikes. We patch arterial tubing with plastic; we salvage corneas from the dead, and kidneys from the living or dead; automatic pacemakers placed in the brain case may shortly control behavior. There are artificial limbs, artificial lungs, artificial kidneys and artificial hearts in preparation. In 1818 the technologic age existed primarily in the excitement of anticipation. Man was ascending, and the only terror was that in the rise he would offend God by assuming too much, reaching too high and coming too close. The scientist was still cast in the role of Prometheus. The tragic irony is not that Mary Shelley's "fantasy" once again has a relevance; the tragedy is that it is no longer fantasy—and that in its realization we no longer identify with Dr. Frankenstein, but with his monster. But that is an overstatement: we identify with both.

It is not irrelevant that there is a confusion in the mind of the average person, when he uses the term "Frankenstein." To the majority of the population, the word refers to the monster created, rather than the doctor creating it. The confusion is not mere ignorance, but represents our ambiguity. It is a product of our double identification. We would revere Dr. Frankenstein if his monster did not remind us so much of ourselves. And so we are ambivalent—identifying with both creator and creature. The fear of the latter produces the specialized kind of anti-technology bias—the Frankenstein Factor—that has entered into discussions about recombinant DNA research.

What can we do about it? Obviously, we cannot—nor should we—return to a pretechnologic age. Technology has elevated man, and there is no going back. "Natural man" is, at any rate, the collaborative creation of nature and man himself. Anti-technology is a self-hatred that we cannot afford and must not indulge. How, then, can we compensate for this irrational element, this ambivalence—the Frankenstein Factor—that intrudes itself into those kinds of biomedical research? Here, I am not talking about the legitimate entry of the public at large into the enterprises of science. We are entering an age where that is inevitable and, in my judgment, desirable. But public accountability must be shielded from the prejudice of ignorance.

When we are aware that there is a special sensitivity to research that casts man as the material of the research as well as the researcher—particularly when cast in high technology that both excites and frightens laymen—there is much that can be done to defuse the anxiety. If we once again re-examine the recombinant DNA debate, we may begin to see how the Frankenstein Factor operates and what we can do about it.

Let me first state that there are serious concerns by serious people about recombinant DNA research in terms of potential slips and failures in the laboratory, particularly in terms of "escape" material. I am not sure that the dangers here are essentially different from those that have always confronted bacteriologic and viral research. I may be wrong. But at any rate, I do not think that this possibility alone triggers the public anxiety. It is merely the rational reason under whose banner unconscious anxieties were mobilized. The DNA issue was capable of capturing the media and exciting continuous public attention, because it triggered the underlying irrational elements that more often than not tip the balance of rational discourse. I believe that the tacit and unexpressed fear is not just of the failure of the research, but of the nature of the research, and even of its potential success. Success here is not visualized in terms of a conquering of disease or the solution to agricultural problems, but of a creation of new life forms. This outcome, somehow or other, threatens our sense of identity, our sense of uniqueness and our sense of primacy among the creatures of the earth.

With hindsight it is apparent that the well meaning, concerned and dedicated scientists who convened the Asilomar conference in February, 1975, managed to do precisely the opposite of what I would prescribe as a means for defusing and discounting the Frankenstein Factor. First off, rather than allow us, the public, to identify with them, they separated "them" from "us." The conference brought together all the experts in the field, separating those who would do the research from those who saw the research as being "done to" or "on" them. In so doing they forced the public to identify with the monster rather than with Dr. Frankenstein. At the same time, while doing so, they devoted the vast majority of the conference time to highly technologic discussion, which further emphasized the in-group nature of the proceedings. The scientists emphasized the awesome and mysterious technology, and in so doing made it un-understandable and alien to the population at large. One of the tragedies in the follow-up Senate Committee hearings was the insistence of some of the scientists in continuing the separation of "them" and "us," even when "them" were distinguished senators.

Congressmen are less likely to be convinced by the critics of science than they are to be frightened by the aloofness, bordering sometimes on arrogance, of its defenders. There is still an enormous reservoir of awe and respect among laymen toward science. As stated explicitly in testimony (and suggested implicitly by the composition of the Asilomar group), the research could only be understood, and therefore could only be controlled, by the experimenters. The risk, however, would be shared by all. Proud in their achievement of halting a major research, they engineered the most successful publicity and promotional package of any scientific meeting that I have ever encountered. Asilomar became a scientific version of "Jaws," and the public was titillated, but also frightened. In generating awe, however, they encouraged the fear that is a component of it. In their pride in their achievement, the researchers forgot that the public would also be aware that they resumed the research some short period later and, more important, that it was "they," the scientists, who both called a halt and allowed the resumption. Precisely the opposite approach was needed.

All research is now publicity-prone—and open for public supervision. But some is more vulnerable to hysteria than others. We must be extraordinarily sensitive to the Frankenstein Factor and aware when it is likely to be mobilized. There are ways to mitigate it. The nature of the technology must be de-mystified. This step requires a special quality of rhetoric that emphasizes the fundamental understandability of purpose rather than the esoteric qualities of technic. Secondly, high technology should be related in its basic principles to less frightening low-technologic proceedings of the same sort. This is simply another way of saying, show the continuity between the new—and therefore threatening—and the old and familiar. There are, after all, similarities between surgical modification of behavior and conditioning, as well as differences. There are also similarities between the projected uses of recombinant DNA research and old, comfortable and now accepted as routine, medical procedures. The continuum must be emphasized with the recognition that the present-day familiar, at its inception, also invoked the anxiety of the unknown and new. And, finally, researchers must be aware of when the Frankenstein Factor is most likely to be invoked. With such cases they must not only allow, but indeed invite, the participation of the public in the decision making apparatus. In so doing they will encourage the identification of the public with those who are doing, rather than those who will be done to. Such participation will ultimately enhance scientific freedom of inquiry. By eliminating the Frankenstein

Factor, we will encourage the rational climate so essential for proper public policy.

Source: Willard Gaylin, "The Frankenstein Factor," *New England Journal of Medicine* 297 (September 22, 1977): 665–667.

DOCUMENT 110: James F. Keenan, "What Is Morally New in Genetic Manipulation?" (1990)

New issues in genetics require us to examine carefully the new ethical dilemmas they raise. But that requires a further ethical examination: a determination of whether our context and method of moral reflection are in fact adequate to address such issues. As noted in the following document, this can be interpreted to mean that a new ethical method—the approach of virtue-based ethics instead of the customary act-orientated ethics—is required to help us resolve the new issues raised by genetics. In addition, the major problem we face is the potential objectification of the human subject.

* * *

THE NEW MORAL CHALLENGE

If we hold that the human person is to be treated as a subject, then we see something morally extraordinary in treating a human person as an object. In fact, the enduring critique of medical work has been whenever the person is not treated as a person but as a body or object. For instance, the calculus used in determining moral licitness of experimentation on human subjects usually concerns a proportionality of the benefits from the research versus the risk of objectifying, dehumanizing, or harming the human subjects. This same calculus would be applicable for judging the licitness of genetic experimentation on human subjects.

In genetic manipulation, the issue of the objectification of the human person permeates the four-fold distinction. In fact, the distinction itself marks the progressive objectification of the human person. First, analogous to present medical practices, therapeutic manipulation objectifies the disease in the person rather than the person. In a similar vein, germ-line gene therapy is distinct from both enhancement and eugenics also in terms of objectification. The object of our concern in germ-line gene therapy is the genetic disorder that we are correcting or eliminating. Nonetheless, inasmuch as we are altering the future genetic material of a future person (one may dispute this, but the very ground for the manipulation is for the person whose genetic material this will be), we are not in the same position to differentiate the disease from the person as we are in gene therapy. Whereas the person and the disease are distinguishable in the first distinction, they are not so in the second.

In enhancement, the human genotype itself becomes the object of our work. We are no longer attempting to objectify a disease for treatment; we are rather objectifying a person for treatment. Rather than simply correcting a particular genetic disorder even in a not-yet-person, we are partially redesigning the genotype itself and therein begin a process of objectifying the potential human subject. In eugenics, the agenda is precisely that objectification. The design is no longer partial; the design is complete and the potential human person is first an object before being a subject.

The newness here is not precisely objectification. In the past, objectification has always been thoroughly immoral, for generally speaking objectification was an act of oppression, slavery being a key manifestation of it. What is new is a complicated insight with two parts: First the biologically programmatic nature of this objectification is not itself necessarily immoral as previous objectifications were, but second, if it were used for immoral purposes its exercise would be more harmful than any other previous objectification. As C. S. Lewis wrote:

"If any one age really attains, by eugenics and scientific education, the power to make its descendants what it pleases, all men who live after it are patients of that power. They are weaker, not stronger: for though we may have put wonderful machines in their hands we have pre-ordained how they are to use them. . . . The real picture is that of one dominant age . . . which resists all previous ages most successfully and dominates all subsequent ages most irresistibly, and thus is the real master of the human species. But even within this master generation (itself an infinitesimal minority of the species) the power will be exercised by a minority smaller still. Man's conquest of Nature, if the dreams of the scientific planners are realized, means the rule of a few hundreds of men over billions upon billions of men." (1)

Whereas in the past, the oppressed at least had the potential to resist, if eugenics were an act of domination, the new victims would not have that same potential. In any case, the progressive objectification of the person will be the enduring new moral dilemma as genetic manipulation expands. (2) No longer will medicine be able to distinguish the disease or health of the person from the person her/himself. And somehow the only morally legitimate foreseeable course of action in each of the distinctions will be the creation of conditions in which the person, though objectified, is not solely treated as an object. This will be the enduring challenge of genetic manipulation.

NOTES

1. Lewis, C. S. (1965). *Abolition of Man*. (Collier-Macmillan, New York) pp. 70–71.

2. See especially McCormick, "Genetic Technology and Our Common Future," and Rahner, "The Experiment with Man."

Source: James F. Keenan, "What Is Morally New in Genetic Manipulation?" *Human Gene Therapy* 1 (1990): 289–298.

DOCUMENT 111: Thomas A. Shannon, "Ethical Issues in Genetic Engineering: A Survey" (1992)

The following document surveys many ethical questions associated with gene therapy. Its main purpose is to highlight these problems to enhance our awareness that the important issues of the new genetics are ethical as well as scientific. Of particular importance are the various motives for engaging in genetic engineering as well as the application of new genetic information to the prenatal diagnosis of various diseases.

* * *

Introduction

Of all the developments in science, medicine and engineering in recent decades, perhaps none has captured the imagination as much as genetic engineering. In one way this is surprising because genetics is not easily understandable, particularly given the complexity of the genetic code.

But in another way the fascination is understandable. For even with minimal understanding of genetics, we can sense the attraction of engineering the engineer. That is, we have the capacity to influence directly the basic units of inheritance and thus shape not only the lives of specific human beings but also the course of evolution itself.

Our imaginations have been captured by literary and film portrayals of the applications of genetic engineering. Novels such as Huxley's *Brave New World* and Robin Cook's *Mutation* present the dark side of genetic engineering with frightening social and political implications. The film *Blade Runner* presents genetically engineered warriors going out of control, as well as an interesting relationship between a human being and a genetically engineered entity. A movie from several years ago, *The Boys From Brazil*, presents an attempt to replicate other Hitlers by cloning and replicating the major developmental events of Hitler's childhood and adolescence.

Since the DNA molecule was first described by Watson and Crick in 1953, researchers have steadily advanced our understanding of human genetics. As a means of coordinating this investigation, scientists have organized the Human Genome Project, one of the largest research programs of the century. Its purpose is to map the entire human genome. This paper will explore some ethical and social implications of emerging genetic technologies.

Genetic Engineering: The Ethical Issues

Motive

Apart from satisfying the seemingly innate curiosity we have about everything in our world, why would we want to study genetics and engage in genetic engineering?

First of all, knowing what information is transmitted genetically will help us identify the cause and mechanism of many diseases that are passed from parent to child. Knowledge about the causes, location and operation of these genetic transmitters will give us the capacity to intervene and replace the faulty gene with a correct working model.

Here the motive of genetic engineering is the traditional medical motive of providing therapy for one who is or may become ill. The difference, and it is a critical one, is that the disease will be prevented from occurring because the gene which causes it will be replaced with a correct copy. Such a possibility would be a blessing for those who are carriers for a particular disease, sickle cell anemia or Huntington's chorea, for example, but would like to have children. This therapeutic motive for genetic engineering stands well within the traditional ethic of medicine and presents the same or similar ethical issues involved in the customary practice of medicine.

A second motive is to use various genetic technologies or interventions to enhance a particular trait or characteristic. A commonplace example, but quite profound both socially and economically, is the commercial use of selective breeding techniques to produce chickens with enhanced body parts but which also require less feed to attain their size and can reach market quicker with a better profit margin.

On a different level, animal breeders are cloning several identical copies of animals from a single embryo. In this technology, the genetic material in the unfertilized eggs of a donor cow is removed and replaced with genetic material from a prize cow. These eggs are then frozen for later use or are fertilized with semen from prize bulls, implanted into surrogates who then give birth to genetically identical calves. Thus the desired traits are more quickly produced and brought to market.

Transgenic animals are also becoming a staple of the laboratory and farm. A transgenic animal is one who has genetic information from another species inserted into it. A successful example has been the development of mice whose milk produces an agent used to prevent heart attacks in humans. Cows have been given a genetically engineered enzyme to enhance their milk production. And, in the opposite direction, a mouse has been engineered to make it susceptible to cancer so it can be used for research.

Here the intent is to enhance a particular capacity or characteristic of the entity in question. And it is here that many of our most difficult ethical and social questions arise. Some have asked whether it is appropriate to use animals in this way. This question is sharpened in light of the possible upset of the ecological balance and the impact on evolutionary development. Another question is whether it is possible for humans to own patents on animals, particularly when that animal has a capacity not found in nature and which was put there by researchers or institutions seeking such a patent. While the legal dimension of this question has been answered affirmatively in the United States, the ethical dimension is being reviewed from the perspective of the developing animal rights movement. Considerable rethinking of this issue is going on at present.

But more urgent questions concern the application of these technologies to human beings. One example is the use of genetically engineered human growth hormone. Originally developed to correct dwarfism, the hormone is now available for children with "short child syndrome," that is, normal children who are of average or less than average height. Since we know that taller people command greater respect and higher salaries, it is not surprising that many parents want to ensure the advantage of height for their children. But short stature is not an illness, thus on what medical grounds might a physician legitimately prescribe it?

Another example is Dr. Robert Graham and his Repository for Germinal Choice. Having made a fortune by inventing safety glasses, Dr. Graham was motivated to help society by increasing the number of highly intelligent people in the expectation that they would become scientists. To do this, he asked Nobel Prize winners to donate their sperm so it could be used in artificial insemination. Dr. Graham neglected the genetic contribution by the woman and the fact that it was the parents of the prize winners who contributed the genetic information. He also incorrectly assumed that there is a single gene responsible for intelligence.

As we develop the capacity to engineer the engineer, we need to keep in mind what we are trying to achieve and ask why that is a good idea.

Prenatal Diagnosis

Another common application of genetic knowledge is the routine screening of pregnancies by amniocentesis, chorionic villi sampling, ultrasound or fetoscopy. In the first two techniques fetal cells are obtained,

cultured and then examined for genetic anomalies. In the last two procedures the embryo or fetus is observed either indirectly via a monitor
or directly with the use of fiber optic cables. These technologies help us
observe developmental progress and discover various anomalies.

While several thousand genetic diseases can be diagnosed in this fashion, the tragedy is that the vast majority cannot be cured. Occasionally
an intervention can be made, such as a blood transfusion or immediately
initiating a particular diet, but in the majority of cases we only gain
knowledge from the diagnosis. Occasionally an anomaly is discovered
whose effect is not known. Here the parents will be in a position of
knowing there is a genetic variation but not knowing what effects, if
any, might follow from it.

Another dimension of genetic screening is learning the sex of the embryo. Such knowledge is medically relevant if, for example, the embryo
is at risk for a sex-linked disease such as hemophilia. But suppose, as is
sometimes the case, that the individual or couple wants to know the sex
because they have a birth order preference for their children? Polls show
that most individuals, including women, prefer sons as the firstborn.
Acting on such knowledge could significantly skew population distri-
butions, would confer additional social advantages on males, and would
introduce sexual discrimination into the birth process. One needs to ask
whether it is legitimate to use a medical test to ascertain information that
is almost exclusively social in nature.

While genetic screening has its place, particularly if a couple who are
not obvious carriers of a genetic disease have an affected child or if either
or both partners is a carrier of a genetic disease, there are several important ethical issues. First, there is a "guilty until proven innocent"
assumption effectively acting here. That is, the fetus is suspected of
having a genetic disease until proven otherwise. Since only about three
percent of fetuses so screened are shown to have anomalies, one wonders
about the legitimacy of such an assumption and why screening is so
prevalent.

Screening assumes that those without genetic anomalies will incur
lower health care costs or utilize fewer of the community's resources.
While it is true that individuals with genetic illnesses do incur health
care costs, it is also true that so does almost everyone else. And the
assumption that such causes are primarily genetic ignores the fact that
many illnesses are caused by environmental pollution as well as by personal choice. For example, the practice of smoking is not caused genetically, but it accounts for a tremendous health care expense annually—
as does drug addiction and drunk driving.

Finally, information about the genetic status of the fetus is typically
gained late in pregnancy, usually after the 15th week. The problem is
that the woman or couple are emotionally invested in the pregnancy by

now, typically a pregnancy of choice or one that was achieved only after great difficulty. A significant amount of tension is introduced into the pregnancy by the assumption that there may be an anomaly, by the two or three week waiting period for the test results, and the difficulty of what decision to make should there be a genetic error. All of this contributes to a new form of pregnancy that Barbara Katz Rothman calls the "tentative pregnancy."

When indications call for it, prenatal diagnosis can be a helpful way of diagnosing problems. But we need to think carefully about the personal and social implications of routine screening.

The Human Genome Project

The first serious discussions of providing a map of the human genome began in the late 1980s and led rather rapidly to the finding of a project to implement the idea. Several factors drive this project. One is the basic curiosity we have about ourselves: who are we, what makes us up, why we act the way we do, and how we learn, remember and experience? Scientists hope that a fuller understanding of the genome will supply insight into these basic questions. Clinical investigators seek specific medical gains. Once one learns where a particular gene is and what it does, one can then intervene to correct a malfunctioning gene. This would open the possibility of therapeutic intervention for many of our most serious diseases. By knowing an individual's genetic structure, one might also be able to make predictions about future health or vulnerability to particular illnesses. Thus the genome map offers significant opportunities for gains in knowledge and therapeutic intervention.

But, as in all other areas of life, these opportunities come with the equally strong possibility of high costs. For example, on the scientific level, commitment to the genome project means that many other scientific projects will not be funded. This reality is causing heated debate among scientists who see many worthy prospects being put on hold for several decades.

Several other thematic issues arise from the genome project. The question of control or ownership of information is critical. Is the map of the genome public property, accessible to any scientist who wants particular genetic information for his or her research? Or will this map be patented and available only for a fee? Will it be the property of the government or of the scientists, laboratories or universities at which the work was done? These issues have not been fully studied, much less resolved.

Who will have access to this information? Obviously, full genetic information about individuals would be of interest to insurance companies or employers who provide health care benefits. But is the knowledge of one's genetic makeup included in the right to privacy?

Will the knowledge we learn about ourselves open the way to a "genetic fatalism" in which we assume that because we have a certain genetic configuration we are destined to live or act in a certain way? Will we assume that learning an individual's genetic composition gives the key insight into him or her, and thereby neglect the environment in which he or she lives or the nurturing he or she receives? Could genetic profiles lead to a social ranking of individuals?

It is clear that some diseases are caused by genes that malfunction or are damaged and that short of replacing the defective gene with a working copy, there is little one can do to prevent the disease from occurring. But is this the model for all human activities or behavior? Again, such issues are not resolved and we would do well to examine carefully claims that all behavior can be traced to specific genes.

Conclusion

This paper has reviewed a few important developments in genetics. Some of these are older in that the technologies have been with us for several decades, whereas others are newer and are a consequence of recent advances. Yet all of these techniques, plus the knowledge they generate, significantly challenge our assumptions about human nature, our values and our understanding of society.

The genetic revolution assumes (as does much of our culture) that information and knowledge will resolve many of our problems, either individual or social. That is, once we have a map of an individual's genetic structure we will know much about his future. We will have a privileged insight into his potential and abilities. We will know what makes him tick.

But will we? While it is clear that our genes are responsible for much of our behavior and our physical dimension, it is not clear that genetic structure provides a full or adequate account of human behavior. While research continues, and it should, we need to be constantly on guard about simplistic explanations of complex behavior.

Many of the discussions about genetic engineering, particularly those in either literary or film science fiction, reveal interesting value dimensions. Most of these works present images of genetically engineered warriors or slaves. The values emphasized range from power, intelligence, strength to docility and obedience. Seldom does one find portrayals of gentleness, kindness, compassion, tenderness, creativity or concern for others. When one reads discussions of what traits to select or which characteristics to enhance, these are definitely not the ones mentioned! What does that say about us both individually and as a society? What does it say about our priorities?

The most important debate in genetics may not be about which project to fund or which technique to implement, though these are surely pivotal. Rather, the most critical debates highlight the values on which we base our decisions and the priorities we are trying to actualize in our society.

The creation account in the book of Genesis tells us that we are created in the image of God. The rapidly developing capacities of genetic engineering will give us the power to create our descendants in another image. Whose image will that be, and what values will it embody? Such is the individual and social debate before us as we enter a new age of genetic discovery.

Source: Thomas A. Shannon, "Ethical Issues in Genetic Engineering: A Survey," *Midwest Medical Ethics*, summer 1992, 26–29.

DOCUMENT 112: John Maddox, "New Genetics Means No New Ethics" (1993)

This article argues, from a European perspective, that new developments in genetics will not present new ethical problems. Rather, the author argues, the problems of the new genetics are similar to the problems of the old genetics and should not cause alarm. The author's method of analysis presents an interesting perspective on the blending of scientific understanding and ethical reasoning.

* * *

These days, everybody seems to have an opinion on the ethics of genetics. And most opinions are portentous, laden with apprehension and downright distrust. The new genetics seems to be as widely feared as were the development and deployment of nuclear weapons in the 1950s. It may not be long before geneticists enjoy the popular reputation that then attached to Stanley Kubrick's celluloid character Dr. Strangelove, and when biotechnology enterprises that have proudly chosen corporate names including "GENE" or "GEN" will be scampering for politically more correct alternatives. But the widespread fear of genetics cannot be justified. On the contrary, the research community should speak out strongly to defend the good sense of what it is about.

Present difficulties go back nearly two decades to the Asilomar conference at which, in 1975, a group of molecular biologists recommended a moratorium on genetic manipulation while arrangements were made to regulate recombinant-DNA techniques. Even with the largely reassuring knowledge that experience has brought, that was a wise decision. There might have been hidden dangers in the innocent manipulation of microorganisms. Moreover, the malevolent manipulation of microorganisms could still make possible the manufacture of dangerous biological agents. That is the case for continuing to regulate genetic manipulation.

Asilomar has nevertheless left an awkward legacy. Without precedent, the research community invited legislation for and regulation of its own activities. Valuable though the outcome has been in making genetic manipulation widely understood and almost as widely acceptable, the precedent is now taken to apply to all novel research in biology. But experience has also shown that governments find it easier to create regulations than to dispense with them when the hypothetical dangers they are designed to avoid are shown to be negligible. Asilomar has become a lightning-rod for generalized anxiety about genetics in any form.

Even so, it is not easy to see why the more recent development of techniques for determining the nucleotide sequences of genes, followed by the ambition to sequence the whole human genome, should be of such interest to the people now called "ethicists." Part of the trouble is the difficulty of explaining the difference between discovery and its application; there is a temptation to believe that the first sequence of a human genome will be quickly followed by the widespread application of that knowledge without material impediment, such as cost. There is also a temptation to believe that new laboratory techniques mean new application and thus new ethical problems. But that is not the case. Indeed, the availability of gene sequences, and ultimately of the sequence of the whole genome, will not create ethical problems that are intrinsically novel, but will simply make it easier, cheaper and more certain to pursue well-established objectives in the breeding of plants, animals and even people. That is the principle by which the research community should stand.

The attention paid in the past few years to the nucleotide sequences of genes involved in the inheritance of genetic diseases is a good example. The molecular causes of conditions such as sickle-cell anemia, the various thalassaemias, Huntington's disease, fragile-X syndrome and cystic fibrosis have all been determined in the past few years. But this new knowledge has not created novel ethical problems, only ethical simplifications.

Long before gene sequencing, it had been recognized that these conditions were genetic: they run in families. It became standard practice to warn putative carriers of disease genes of the risks of conceiving children. Is it not a welcome improvement of that practice that genetic counselors no longer have to rely on inference, based on imperfect family histories, in giving their advice? Or that when the impending birth of a child with a genetic disease can be determined by amniocentesis, that most governments should allow abortion?

But may not such practices be carried too far? One fear is that the avoidance of births of individuals carrying genes supposed "defective," but especially the elimination of those genes from a population, may unknowingly rob society of valuable genetic endowments. The heterozygous advantage (resistance to malaria) of the sickle-cell gene is the classical example. The possibility that the "gene" for schizophrenia, which must be commonplace if it exists, may confer advantages not yet recognized on those in whom the overt disease does not develop is often cited as a more serious pitfall.

The truth is not simply the practical consideration that it will be a long time before the genetics of these psychiatric conditions is understood, but that geneticists themselves are likely to be the first to recognize the dangers of interfering with the natural flow of genes within a population before the social implications are understood. Indeed, only geneticists can recognize the dangers. More general recognition of them should help enormously to modify for the better the now too-common view that genes spell genetic trouble whenever they appear in a non-standard form. (On the other hand, to refrain from calling the Huntington's gene "abnormal" is to concede too much to political correctness.)

There are more subtle dangers. Because the nucleotide sequence of a person's genome makes it possible to discriminate between people, genetic sequencing could become the basis for refusing individual medical insurance or even employment in some occupations. All that is true in principle, but there are three good reasons why the threat does not amount to a novel ethical challenge.

First, insurers already discriminate against people who smoke cigarettes, or who are infected with HIV; why should they not also discriminate against, say, people with the particular structure of the LDL receptor known to be responsible for early-onset familial heart disease? Second, would it not be beneficial that people carrying the heart disease gene should not be employed on heavy manual work (or better, that they should be steered towards drug treatment)? Third, there are practical problems: when 100,000 genes have been sequenced, how many will insurers or employers include in their genetic screens, and at what cost?

Then there is the matter of eugenics. Genetic screening by amniocentesis does not decrease the frequency of an unwanted recessive gene in a population but may even increase it. And of necessity, gene therapy designed to improve on the defective function of some class of somatic cells can have no effect on the condition of others than the individual patient. But what if it were possible (as it is not, yet) to modify germ cells in the human reproductive system?

Reference to Hitler is common at this stage, but is mistaken. What serious geneticist would at this stage advocate the artificial selection of a particular genome, throwing away the benefits of hybrid vigour and genetic diversity? But would it not be of great value if the frequency of the haemophilia gene in the human population could somehow be reduced? Even if (when a safe way of doing that has been found) by germ-cell manipulation? Geneticists are fond of saying they will "never touch the germline." That is unwise.

Source: John Maddox, "New Genetics Means No New Ethics," *Nature* 364 (July 8, 1993): 97.

DOCUMENT 113: Executive Summary of the NIH-DOE Working Group on Ethical, Legal, and Social Implications of Human Genome Research (1993)

This document suggests a variety of ways to deal with the health implications of the new genetic information that bombards us, health care professionals, and insurance companies daily. The key problem is the use of such genetic information and whether or not medical coverage can be denied based on genetic information an insurance company finds unacceptable. Genetic discrimination is a possibility, but is it a preventable possibility?

* * *

One of the ironies in the current health care coverage crisis is that developing more accurate biomedical information could make things worse rather than better.

In the current American health care system, information about an individual's risk of disease plays a crucial role for many people in determining access to health care coverage. This link between the likelihood of needing health care and the ability to obtain coverage for that care has the unfortunate result that those most in need may have the greatest difficulty finding affordable health care coverage. New advances in human genetics are transforming medicine by making available increasing amounts of such information about risk.

Biomedical science and the delivery of health care are being reshaped by advances in our understanding of human genetics. New insights into health and disease, new diagnostic and prognostic tests and the possibility of new therapies reflect significant investments by the public and by private business and are no longer limited to the uncommon disorders traditionally labeled as "genetic diseases." Among the first products of genetic research is information useful in predicting the likelihood that an individual will develop particular diseases, opening the door both to preventive strategies that we would welcome, such as changes in diet and exercise patterns, and to the unwelcome possibility of genetic discrimination.

Injecting information about genetic risks into the current health care system could result in ever more refined risk rating by insurers and ever greater difficulty in finding affordable health care coverage for large numbers of people. At a minimum, people could be discouraged from obtaining genetic information that might be useful in disease prevention and early treatment or for case planning and management because that same information could jeopardize their access to health care coverage in general, or to treatment for a condition excluded from coverage because it was "preexisting." Under other circumstances people might be compelled to provide genetic information as a condition of obtaining affordable health care coverage. Genetic risk information carries an additional, wider burden because information about an individual's genetic health risks may also be information about the risks of children, parents, brothers, sisters, and other relatives.

One suggested approach—providing special protection for genetic information—is unlikely to succeed. This special protection has been suggested because of the relevance of genetic information to family members and its implications for reproductive choices, potential discrimination and stigmatization. Genetic privacy ought to be vigorously protected; however, other varieties of health related information are equally sensitive. Furthermore, as a practical matter, genetic information is not segregated from other health related information in, for example, medical records.

Special protection for genetic information is also difficult to enforce because of the "genetic revolution" in medicine. Diseases increasingly are coming to be seen as having both genetic and non-genetic components, making it ever more difficult to classify health related information as wholly genetic or non-genetic. The standard personal medical history, for example, is a rich source of genetic information. Policies intended to protect genetic privacy will need to address the privacy of health related information in general. If we want strict standards to safeguard genetic information, then those same standards will have to extend to all health related information. The Task Force considered these factors carefully.

In anticipation of fundamental reform in the financing and delivery of health care in the U.S., the Task Force on Genetic Information and Insurance offers the following recommendations. The recommendations concern health care coverage and should not be applied uncritically to other forms of insurance, such as life or disability income insurance.

—Information about past, present or future health status, including genetic information, should not be used to deny health care coverage or services to anyone.

—The U.S. health care system should ensure universal access to and participation by all in a program of basic health services* that encompasses a continuum of services appropriate for the healthy to the seriously ill.

—The program of basic health services should treat genetic services comparably to non-genetic services, and should encompass appropriate genetic counseling, testing and treatment within a program of primary, preventive and specialty health care services for individuals and families with genetic disorders and those at risk of genetic disease.

—The cost of health care coverage borne by individuals and families for the program of basic health services should not be affected by information, including genetic information, about an individual's past, present or future health status.

—Participation in and access to the program of basic health services should not depend on employment.

—Participation in and access to the program of basic health services should not be conditioned on disclosure by individual and families of information, including genetic information, about past, present or future health status.

—Until participation in a program of basic health services is universal, alternative means of reducing the risk of genetic discrimination should be developed. As one step, health insurers should consider a moratorium on the use of genetic tests in underwriting. In addition, insurers could undertake vigorous educational efforts within the industry to improve the understanding of genetic information.

(We use the phrase "program of basic health services" to describe the array of services that would be available to all after implementation of major health policy reforms. We explicitly reject all connotations of "basic" as minimal, stingy, or limited to such services as immunization and well child care. A program of "basic" health services would encompass a broad range of care for those most in need.)

Source: Executive Summary, NIH-DOE Working Group on Ethical, Legal, and Social Implications of Human Genome Research (1993).

DOCUMENT 114: *Diamond v. Chakrabarty* (1980)

This document presents the ethical reasoning of the Supreme Court of the United States, which decided that a living organism that is artificially created in the laboratory may be patented. The organism under discussion is a genetically engineered bacterium that breaks down oil and is designed to be used in cleaning up oil spills. The critical finding of the Court is that the living status of this organism confers no special status with respect to patent law.

* * *

Title 35 U.S.C. §101 provides for the issuance of a patent to a person who invents or discovers "any" new and useful "manufacture" or "composition of matter." Respondent held a patent application relating to his invention of a human-made, genetically engineered bacterium capable of breaking down crude oil, a property which is possessed by no naturally occurring bacteria. A patent examiner's rejection of the patent application's claims for the new bacteria was affirmed by the Patent Office Board of Appeals on the ground that living things are not patentable subject matter under §101. The Court of Customs and Patent Appeals reversed, concluding that the fact that micro-organisms are alive is without legal significance for purposes of the patent law.

Held: A live, human-made micro-organism is patentable subject matter under §101. Respondent's micro-organism constitutes a "manufacture" or "composition of matter" within that statute. Pp. 308–318.

(a) In choosing such expansive terms as "manufacture" and "composition of matter," modified by the comprehensive "any," Congress contemplated that the patent laws should be given wide scope, and the relevant legislative history also supports a broad construction. While laws of nature, physical phenomena, and abstract ideas are not patentable, respondent's claim is not to a hitherto unknown natural phenomenon, but to a nonnaturally occurring manufacture or composition of matter—a product of human ingenuity "having a distinctive name, character [and] use." *Hartranft v. Wiegmann*, 121 U.S. 609, 615. *Funk Brothers Seed Co. v. Kalo Inoculant Co.*, 333 U.S. 127, distinguished. Pp. 308–310.

(b) The passage of the 1930 Plant Patent Act, which afforded patent protection to certain asexually reproduced plants, and the 1970 Plant Variety Protection Act, which authorized protection for certain sexually reproduced plants but excluded bacteria from its protection, does not evidence congressional understanding that the terms "manufacture" or "composition of matter" in §101 do not include living things. Pp. 310–314.

(c) Nor does the fact that genetic technology was unforeseen when Congress enacted §101 require the conclusion that micro-organisms cannot qualify as patentable subject matter until Congress expressly authorizes such protection. The unambiguous language of §101 fairly embraces respondent's invention. Arguments against patentability under §101, based on potential hazards that may be generated by genetic research, should be addressed to the Congress and the Executive, not to the Judiciary. Pp. 314–318.

MR. CHIEF JUSTICE BURGER delivered the opinion of the Court.

We granted certiorari to determine whether a live, human-made micro-organism is patentable subject matter under 35 U.S.C. §101.

I. In 1972, respondent Chakrabarty, a microbiologist, filed a patent application, assigned to the General Electric Co. The application asserted 36 claims related to Chakrabarty's invention of "a bacterium from the genus Pseudomonas containing therein at least two stable energy-generating plasmids, each of said plasmids providing a separate hydrocarbon degradative pathway." (1) This human-made, genetically engineered bacterium is capable of breaking down multiple components of crude oil. Because of this property, which is possessed by no naturally occurring bacteria, Chakrabarty's invention is believed to have significant value for the treatment of oil spills. (2)

Chakrabarty's patent claims were of three types: first, process claims for the method of producing the bacteria [p. 306]; second, claims for an inoculum comprised of a carrier material floating on water, such as straw, and the new bacteria; and third, claims to the bacteria themselves. The patent examiner allowed the claims falling into the first two categories, but rejected claims for the bacteria. His decision rested on two grounds: (1) that micro-organisms are "products of nature," and (2) that as living things they are not patentable subject matter under 35 U.S.C. §101.

Chakrabarty appealed the rejection of these claims to the Patent Office Board of Appeals, and the Board affirmed the examiner on the second ground. (3) Relying on the legislative history of the 1930 Plant Patent Act, in which Congress extended patent protection to certain asexually reproduced plants, the Board concluded that §101 was not intended to cover living things such as these laboratory created micro-organisms. The Court of Customs and Patent Appeals, by a divided vote, reversed on the authority of its prior decision in *In re Bergy*, 563 F.2d 1031, 1038 (1977), which held that "the fact that microorganisms . . . are alive . . . [is] without legal significance" for purposes of the patent law. (4) Subsequently, we granted the Acting Commissioner of Patents and Trademarks' petition for certiorari in *Bergy* vacated the judgment, and remanded the case "for further consideration in light of *Parker v. Flook*, 437 U.S. 584 (1978)." 438 U.S. 902 (1978).

The Court of Customs and Patent Appeals then vacated its judgment in *Chakrabarty* and consolidated the case with *Bergy* for reconsideration. After re-examining both cases in the light of our holding in *Flook* that court, with one dissent, reaffirmed its earlier judgments. 596 F.2d 952 (1979).

[p. 307] The Commissioner of Patents and Trademarks again sought certiorari, and we granted the writ as to both *Bergy* and *Chakrabarty*. 444

U.S. 924 (1979). Since then, Bergy has been dismissed as moot, 444 U.S. 1028 (1980), leaving only *Chakrabarty* for decision.

II. The Constitution grants Congress broad power to legislate to "promote the Progress of Science and useful Arts, by securing for limited Times to Authors and Inventors the exclusive Right to their respective Writings and Discoveries." Art. 1, §8, c. 8. The patent laws promote this progress by offering inventors exclusive rights for a limited period as an incentive for their inventiveness and research efforts. *Kewanee Oil Co. v. Bicron Corp.*, 416 U.S. 470, 480–481 (1974); *Universal Oil Co. v. Globe Co.*, 322 U.S. 471, 484 (1944). The authority of Congress is exercised in the hope that "[the] productive effort thereby fostered will have a positive effect on society through the introduction of new products and processes of manufacture into the economy, and the emanations by way of increased employment and better lives for our citizens." *Kewanee*, supra, at 480.

The question before us in this case is a narrow one of statutory interpretation requiring us to construe 35 U.S.C. §101, which provides:

"Whoever invents or discovers any new and useful process, machine, manufacture, or composition of matter, or any new and useful improvement thereof, may obtain a patent therefore, subject to the conditions and requirements of this title."

Specifically, we must determine whether respondent's micro-organism constitutes a "manufacture" or "composition of matter" within the meaning of the statute. (5)

III. In cases of statutory construction we begin, of course, with the language of the statute. *Southeastern Community College v. Davis*, 442 U.S. 397, 405 (1979). And "unless otherwise defined, words will be interpreted as taking their ordinary, contemporary, common meaning." *Perrin v. United States*, 444 U.S. 37, 42 (1979). We have also cautioned that courts "should not read into the patent laws limitations and conditions which the legislature has not expressed." *United States v. Dubilier Condenser Corp.*, 289 U.S. 178, 199 (1933).

Guided by these canons of construction, this Court has read the term "manufacture" in §101 in accordance with its dictionary definition to mean "the production of articles for use from raw or prepared materials by giving to these materials new forms, qualities, properties, or combinations, whether by hand-labor or by machinery." *American Fruit Growers, Inc. v. Brogdex Co.* 283 U.S. 1, 11 (1931). Similarly, "composition of matter" has been construed consistent with its common usage to include "all compositions of two or more substances and . . . all composite articles, whether they be the results of chemical union, or of mechanical mixture or whether they be gases, fluids, powders or solids." *Shell Development Co. v. Watson*, 149 F.Supp. 279, 280 (DC 1957) (citing 1 A. Deller, Walker on Patents §14, p. 55 (1st ed. 1937)). In choosing such expansive terms as "manufacture" and "composition of matter," modified by the comprehensive "any," Congress plainly contemplated that the patent laws would be given wide scope.

The relevant legislative history also supports a broad construction. The Patent Act of 1793, authored by Thomas Jefferson, defined statutory subject matter as "any new and useful art, machine, manufacture, or composition of matter, or any new or useful improvement [thereof]." Act of Feb. 21, 1793, §1, 1 Stat. 319. The Act embodied Jefferson's philosophy that "ingenuity should receive a liberal encouragement." [p. 309] 5 Writings of Thomas Jefferson 75–76 (Washington ed. 1871). See *Graham v. John Deere Co.*, 383 U.S. 1, 7–10 (1966). Subsequent patent statutes in 1836,1870, and 1874 employed this same broad language. In 1952, when the patent laws were recodified, Congress replaced the word "art" with "process," but otherwise left Jefferson's language intact. The Committee Reports accompanying the 1952 Act inform us that Congress intended statutory subject matter to "include anything under the sun that is made by man." S. Rep. No. 1979, 82nd Cong., 2nd Sess., 5 (1952); H. R. Rep. No. 1923, 82nd Cong., 2nd Sess., 6 (1952). (6) This is not to suggest that §101 has no limits or that it embraces every discovery. The laws of nature, physical phenomena, and abstract ideas have been held not patentable. See *Parker v. Flook*, 437 U.S. 584 (1978), *Gottschalk v. Benson*, 409 U.S. 63, 67 (1972); *Funk Brothers Seed Co. v. Kalo Inoculant Co.*, 333 U.S.127 (1948); *O'Reilly v. Morse*, 15 How. 62, 112–121 (1854); *Le Roy v. Tatham*, 14 How. 156, 175 (1853). Thus, a new mineral discovered in the earth or a new plant found in the wild is not patentable subject matter. Likewise, Einstein could not patent his celebrated law that $E = MC^2$, nor could Newton have patented the law of gravity. Such discoveries are "manifestations of . . . nature, free to all men and reserved exclusively to none." *Funk*, supra, at 130.

Judged in this light, respondent's micro-organism plainly qualifies as patentable subject matter. His claim is not to a hitherto unknown natural phenomenon, but to a nonnaturally occurring manufacture or composition of matter—a product of human ingenuity "having a distinctive name, character [and] [p. 310] use." *Hartranft v. Wiegmann*, 121 U.S. 609, 615 (1887). The point is underscored dramatically by comparison of the invention here with that in *Funk*. There, the patentee had discovered that there existed in nature certain species of root-nodule bacteria which did not exert a mutually inhibitive effect on each other. He used that discovery to produce a mixed culture capable of inoculating the seeds of leguminous plants. Concluding that the patentee had discovered "only some of the handiwork of nature," the Court ruled the product nonpatentable.

Source: U.S. Supreme Court, *Diamond v. Chakrabarty*, 447 U.S. 303 (1980).

DOCUMENT 115: Martin Khor, "A Worldwide Fight against Biopiracy and Patents on Life" (1995)

While this Supreme Court decision clarified the technicalities of what may or may not be patented, others argue against the patenting of plants and animals. The following documents present some of these objections.

* * *

Worldwide opposition to biological piracy is rapidly building up as more and more groups and people become aware that big corporations are reaping massive profits from using the knowledge and biological resources of Third World Communities.

There is growing public outrage that these companies are being treated patents for products and technologies that make use of the genetic materials, plants and other biological resources that have long been identified, developed and used by farmers and indigenous peoples, mainly in countries of the South.

Whilst the corporations stand to make huge revenues from this process, the local communities are unrewarded and in fact the threat in the future is having to buy the products of these companies at high prices.

The transnational corporations are racing one another to manufacture pharmaceutical and agricultural products, the main ingredients of which are the genetic materials of the medicinal plants and food crops of these local communities. The firms are also collecting other living things, ranging from soil microorganisms to animals and the genes of indigenous people, which they use for research and making new products.

These companies are rushing to apply to patent the new products containing the collected genetic materials, so as to prevent competitors from using them. They can then reap larger profits from being able to hike up prices for the products, or by charging royalties to other firms wishing to use the technology.

Source: Martin Khor, Director of the Third World Network, "A Worldwide Fight against Biopiracy and Patents on Life," Third World Network, found at http://www.capside.org.sg/south,twn/title/pat-ch.html. (1995).

DOCUMENT 116: Mark Sagoff, "Should We Allow the Patenting of Life?" (1991)

Traditional animal breeders, like traditional plant breeders, therefore can claim no intellectual property rights in their creations because they have no new knowledge or ideas to contribute to the common inventory of knowledge. Further, makers of transgenic animals cannot describe the design of the organisms they produce in a way that will allow others to manufacture them.

Accordingly, novel varieties of animals have always been considered unpatentable because they are products of skill rather than invention. While novel processes for producing new varieties may be patented, the varieties themselves should not.

Source: Mark Sagoff, Director of the Institute for Philosophy and Public Policy at the University of Maryland, "Should We Allow the Patenting of Life?" *AG Bioethics Forum* (Iowa State University) 3 (August 1991): 4.

DOCUMENT 117: Charles Magel, Bill of Rights of Animals (1998)

2. All animals are entitled to respect. Humanity as an animal species shall not arrogate to itself the right to exterminate or exploit other species. It is humanity's duty to use its knowledge for the welfare of animals. All animals have the right to the attention, care, and protection of humanity.

3. No animals shall be ill-treated or be subject to cruel acts.

4. All wild animals have the right to liberty in their natural environment, whether land, air, or water, and should be allowed to procreate. Deprivation of freedom, even for educational purposes, is an infringement of this right.

7. Animal experimentation involving physical or psychological suffering is incompatible with the rights of animals, whether it be for scientific, medical, commercial, or any other form of research. Replacements must be used and developed.

Source: Charles Magel, *Keyguide to Information Sources in Animal Rights* (Jefferson, N.C.: McFarland, 1998), 233–234.

**DOCUMENT 118: President's Commission for the Study of
Ethical Problems in Medicine and Biomedical and Behavioral
Research,** *Splicing Life: The Social and Ethical Issues of Genetic
Engineering with Human Beings* **(1982)**

This document presents a summary of the findings of a national commission on bioethics on several issues in genetic engineering. This
commission, composed of individuals from many backgrounds and
professions, produced several such studies. Although written in 1982,
this document is still widely cited in the bioethics literature. Although
at the time of its writing many of the issues discussed in this reader
were hypothetical, this document provides a helpful framework for
analysis and evaluation.

* * *

Summary of Conclusions and Recommendations

This Report addresses some of the major ethical and social implications
of biologists' newly gained ability to manipulate—indeed, literally to
splice together—the material that is responsible for the different forms
of life on earth. The Commission began this study because of an urgent
concern expressed to the President that no governmental body was "ex-
ercising adequate oversight or control, nor addressing the fundamental
ethical questions" of these techniques, known collectively as "genetic
engineering," particularly as they might be applied directly to human
beings.

When it first examined the question of governmental activity in this
area, in the summer of 1980, the Commission found that this concern
was well founded. Not only was no single agency charged with exploring this field but a number of the agencies that would have been expected to be involved with aspects of the subject were unprepared to
deal with it, and the Federal interagency body set up to coordinate the
field was not offering any continuing leadership. Two years later, possibly because of the Commission's attention, it appears the Federal agencies are more aware of, and are beginning to deal with, questions arising
from genetic engineering, although their efforts primarily address the
agricultural, industrial, and pharmaceutical uses of gene splicing rather
than its diagnostic and therapeutic uses in human beings.

The Commission did not restrict its examination of the subject to the
responses of Federal agencies, however, because it perceived more important issues of substance behind the expressed concern about the lack
of Federal oversight. The Commission chose, therefore, to address these
underlying issues, although certainly not to dispose of them. On many

points, the Commission sees its contribution as stimulating thoughtful, long-term discussion rather than truncating such thinking with premature conclusions.

This study, undertaken within the time limitations imposed by the Commission's authorizing statute, is seen by the Commission as a first step in what ought to be a continuing public examination of the emerging questions posed by developments and prospects in the human applications of molecular genetics. First, the report attempts to clarify concerns about genetic engineering and to provide technical background intended to increase public understanding of the capabilities and potential of the technique. Next, it evaluates the issues of concern in ways meaningful for public policy, and analyzes the need for an oversight mechanism.

To summarize, in this initial study the Commission finds that: (1) Although public concern about gene splicing arose in the context of laboratory research with microorganisms, it seemed to reflect a deeper anxiety that work in this field might remake human beings, like Dr. Frankenstein's monster. These concerns seem to the Commission to be exaggerated. It is true that genetic engineering techniques are not only a powerful new tool for manipulating nature—including means of curing human illness—but also a challenge to some deeply held feelings about the meaning of being human and of family lineage. But as a product of human investigation and ingenuity, the new knowledge is a celebration of human creativity, and the new powers are a reminder of human obligations to act responsibly.

(2) Genetic engineering techniques are advancing very rapidly. Two breakthroughs in animal experiments during 1981 and 1982, for example, bring human applications of gene splicing closer: in one, genetic defects have been corrected in fruit flies; in another, artificially inserted genes have functioned in succeeding generations of mammals.

(3) Genetic engineering techniques are already demonstrating their great potential value for human well-being. The aid that these new developments may provide in the relief of human suffering is an ethical reason for encouraging them.

—Although the initial benefits to human health involve pharmaceutical applications of the techniques, direct diagnostic and therapeutic uses are being tested and some are already in use. Those called upon to review such research with human subjects, such as local Institutional Review Boards, should be assured of access to expert advice on any special risks or uncertainties presented by particular types of genetic engineering.

—Use of the new techniques in genetic screening will magnify the ethical considerations already seen in that field because they will allow a larger number of diseases to be detected before clinical symptoms are manifest and because the ability to identify a much wider range of genetic traits and conditions will greatly enlarge the demand for, and even the objectives of, prenatal diagnosis.

(4) Many human uses of genetic engineering resemble accepted forms of diagnosis and treatment employing other techniques. The novelty of gene splicing ought not to erect any automatic impediment to its use but rather should provoke thoughtful analysis.

—Especially close scrutiny is appropriate for any procedures that would create inheritable genetic changes; such interventions differ from prior medical interventions that have not altered the genes passed on to patients' offspring.

—Interventions aimed at enhancing "normal" people, as opposed to remedying recognized genetic defects, are also problematic, especially since distinguishing "medical treatment" from "nonmedical enhancement" is a very subjective matter; the difficulty of drawing a line suggests the danger of drifting toward attempts to "perfect" human beings once the door of "enhancement" is opened.

(5) Questions about the propriety of gene splicing are sometimes phrased as objections to people "playing God." The Commission is not persuaded that the scientific procedures in question are inherently inappropriate for human use. It does believe, nevertheless, that objections of this sort, which are strongly felt by many people, deserve serious attention and that they serve as a valuable reminder that great powers imply great responsibility. If beneficial rather than catastrophic consequences are to flow from the use of "God-like" powers, an unusual degree of care will be needed with novel applications.

(6) The generally very reassuring results of laboratory safety measures have led to a relaxation of the rules governing gene splicing research that were established when there was widespread concern about the potential risks of the research. The lack of definitive proof of danger or its absence has meant that the outcome of whether to restrict certain research has turned on which side is assigned the burden of proving its case. Today those regulating gene splicing research operate from the assumption that most such research is safe, when conducted according to normal scientific standards; those opposing that position face the task of proving otherwise.

—The safety issue will arise in a wider context as gene splicing is employed in manufacturing, in agriculture and other activities in the general environment, and in medical treatment. As a matter of prudence, such initial steps should be accompanied by renewed attention to the issue of risk (and by continued research on that subject).

—Efforts to educate the newly exposed population to the appropriate precautions, whenever required, and serious efforts to monitor the new settings (since greater exposure increases the opportunity to detect low-frequency events) should be encouraged. In general, the questions of safety concerning gene splicing should not be viewed any differently than comparable issues presented by other scientific and commercial activities.

(7) The Recombinant DNA Advisory Committee (RAC) at the National Institutes of Health has been the lead Federal agency in genetic engineering. Its guidelines for laboratory research have evolved over the past seven years in response to changes in scientific attitudes and knowledge about the risks of different types of genetic engineering. The time has now come to broaden the area under scrutiny to include issues raised by the intended uses of the technique rather than solely the unintended exposure from laboratory experiments.

—It would also be desirable for this "next generation" RAC to be independent of Federal funding bodies such as NIH, which is the major Federal sponsor of gene splicing research, to avoid any real or perceived conflict of interest.

(8) The process of scrutiny should involve a range of participants with different backgrounds, not only the Congress and Executive Branch agencies but also scientific and academic associations, industrial and commercial groups, ethicists, lawyers, religious and educational leaders, and members of the general public.

—Several formats deserve consideration, including initial reliance on voluntary bodies of mixed public-private membership. Alternatively, the task could be assigned to this Commission's successor, as one among a variety of issues in medicine and research before such a body, or to a commission concerned solely with gene splicing.

—Whatever format is chosen, the group should be broadly based and not dominated by geneticists or other scientists, although it should be able to turn to experts to advise it on the laboratory, agricultural, environmental, industrial, pharmaceutical, and human uses of the technology as well as on international scientific and legal controls. Means for direct liaison with the government departments and agencies involved in this field will also be needed.

(9) The need for an appropriate oversight body is based upon the profound nature of the implications of gene splicing as applied to human beings, not upon any immediate threat of harm. Just as it is necessary to run risks and to accept change in order to reap the benefits of scientific progress, it is also desirable that society have means of providing its "informed consent," based upon reasonable assurances that risks have been minimized and that changes will occur within an acceptable range.

Source: President's Commission for the Study of Ethical Problems in Medicine and Biomedical and Behavioral Research, *Splicing Life: The Social and Ethical Issues of Genetic Engineering with Human Beings* (Washington, D.C.: U.S. Government Printing Office, 1982).

Part VIII

Cloning

After many claims that mammals could not be cloned and after years of laboratory research to back up such claims, history was made on November 27, 1997, when Ian Wilmut and his team of Scottish researchers revealed that they had in fact cloned a ewe, the now famous Dolly. Publication of this claim set forth a torrent of commentaries and pundits. Scientists and ethicists made the round of talk shows—as well as many comedians. Jokes and fantasies eventually gave way to serious discussions, and many countries and religious organizations proposed various guidelines on human cloning—if not recommending outright condemnation. This section presents a sampling of some of the commentaries engendered by the birth of Dolly.

DOCUMENT 119: James D. Watson, "Moving toward the Clonal Man" (1971)

Discussion of cloning did not begin with Dolly. An earlier discussion occurred in the 1970s. The following document presents the early views of the scientist who was the co-discoverer of the DNA molecule—which began the age of modern genetics.

* * *

I would thus hope that over the next decade wide-reaching discussion would occur, at the informal as well as formal legislative level, about the manifold problems which are bound to arise if test-tube conception becomes a common occurrence. A blanket declaration of the worldwide illegalization of human cloning might be one result of a serious effort to ask the world in which direction it wished to move. Admittedly the vast effort, required for even the most limited international agreement, will turn off some people—those who believe the matter is of marginal importance now, and that it is a red herring designed to take our minds off our callous attitudes toward war, poverty, and racial prejudice. But if we do not think about it now, the possibility of our having a free choice will one day suddenly be gone.

Source: James D. Watson, "Moving toward the Clonal Man: Is This What We Want?" *The Atlantic*, May 1971, 50–53.

DOCUMENT 120: Ian Wilmut and others, "Viable Offspring Derived from Fetal and Adult Mammalian Cells" (1997)

The following document is a brief excerpt from the original article announcing the birth of Dolly. It makes the claim of the originality of the process and the critical observation that fully differentiated adult cells can be restarted and become the source of a new being, genetically identical to the donor.

* * *

The lamb born after nuclear transfer from a mammary gland cell is, to our knowledge, the first mammal to develop from a cell derived from an adult tissue. The phenotype of the donor cell is unknown. The primary culture contains mainly mammary epithelial cells (over 905) as well as other differentiated cell types, including myoepithelial cells and fibroblasts. We cannot exclude the possibility that there is a small proportion of relatively undifferentiated stem cells able to support regeneration of the mammary gland during pregnancy. Birth of the lamb shows that during the development of that mammary cell there was no irreversible modification of genetic information required for development to term.

Source: Ian Wilmut and others, "Viable Offspring Derived from Fetal and Adult Mammalian Cells," *Nature* 385 (February 27, 1997): 810–813.

DOCUMENT 121: Vittorio Sgaramella and Norton D. Zinder, "Dolly Confirmation" (1998)

Though not rejecting the significance of Wilmut's claims or denying the experiment, some scientists felt that a number of issues needed to be clarified. Two of these scientists note some of these issues in the document that follows. In particular, they note that the experiment has yet to be replicated, a critical element in the verification of an experiment.

* * *

The single observation gains some credence when well controlled or of a unique nature, or both. It is the lack of any confirmation that provokes our skepticism: here are some of the detailed reasons.

1. The cloning was done once out of some 400 tries. Only one successful attempt out of some 400 is an anecdote, not a result.

2. The characterization of the mammary gland cells used as nucleus donors was poor; it could have been one of the donor's rare stem cells that was involved, as acknowledged in the paper. . . .

5. An analysis of Dolly's mitochondrial DNA has not been given. . . .

7. No hint is given in the paper that the donor ewe had apparently died a few years ago, thereby precluding pertinent immunological testing of genetic relationships.

Source: Vittorio Sgaramella and Norton D. Zinder, "Dolly Confirmation," *Science* 279 (January 30, 1998): 635–666.

DOCUMENT 122: Michael Specter with Gina Kolata, "After Decades and Many Mishaps, Cloning Success" (1997)

Even with such scientific doubts as the ones outlined in the previous document, the success of the cloning experiment set off a major debate over the ethics of the technique. The following document shows that Dr. Wilmut was aware of the social and ethical implications of his research. Yet he helped resolve conflicts by arguing that this technique was not to be used on humans.

* * *

But Dr. Ian Wilmut, the 52-year-old embryologist who astonished the world on Feb. 22 by announcing that he had created the first animal cloned from an adult—a lamb named Dolly—seems almost oblivious to the profound and disquieting implications of his work. Perhaps no achievement in modern biology promises to solve more problems than the possibility of regular, successful genetic manipulation. But certainly none carries a more ominous burden of fear and misunderstanding.

"I am not a fool," Dr. Wilmut said.... "I know what is bothering people about all this. I understand why the world is suddenly at my door. But this is my work. It has always been my work, and it doesn't have anything to do with creating copies of human beings. I am not haunted by what I do, if that is what you want to know."

Source: Michael Specter with Gina Kolata, "After Decades and Many Mishaps, Cloning Success," *New York Times*, March 13, 1997, A1.

DOCUMENT 123: Gina Kolata, "On Cloning Humans, 'Never' Turns Swiftly into 'Why Not?' " (1997)

In spite of the words of Dr. Wilmut, others think that cloning humans could be a possibility. This document, published nine months after the announcement of Dolly, shows how quickly the debate changed.

* * *

In the hubbub that ensued, scientist after scientist and ethicist after ethicist declared that Dolly should not conjure up fears of a Brave New

World. There would be no interest in using the technology to clone people, they said.

They are already being proved wrong.

Some infertility centers that said last spring they would never clone now say they are considering it. A handful of fertility centers are conducting experiments with human eggs that lay the groundwork for cloning. Moreover, the Federal Government is supporting new research on the cloning of monkeys, encouraging scientists to perfect techniques that could easily be transferred to humans. Ultimately, scientists expect cloning to be combined with genetic enhancement, adding genes to give desired traits, which was the fundamental reason cloning was studied in animal research.

Source: Gina Kolata, "On Cloning Humans, 'Never' Turns Swiftly into 'Why Not?' " *New York Times*, December 2, 1997, Al.

DOCUMENT 124: Gina Kolata, "In Big Advance in Cloning, Biologists Clone 50 Mice" (1998)

Finally, on the scientific side, a recent breakthrough in cloning technology was announced by Dr. Ryuzo Yanagimachi of the University of Hawaii, who announced that he had cloned 50 mice, some of whom were clones of clones. Although a different technique was used for cloning the mice—long thought to be impossible to clone because of the rapid development of the embryo—this shows that Dolly was not a lucky accident and that the technology is developing rapidly.

* * *

Dr. Yanagimachi and his colleagues injected the cumulus cell's genetic material into mouse eggs whose own DNA had been removed. They waited six hours to give the egg a chance to reprogram the cumulus cell's DNA and they chemically prodded the egg to start dividing.

Dr. Yanagimachi and his colleagues verified with genetic tests that their mice were indeed clones. Also, in one experiment, they used coffee colored mice for the cumulus cells, black mice for the eggs, and white albino mice as surrogate mothers. As predicted the clones were coffee colored.

Source: Gina Kolata, "In Big Advance in Cloning, Biologists Clone 50 Mice," *New York Times*, July 23, 1998, A1 and A20.

The debate over the ethics of cloning quickly heated up, and many voices provided a wide range of opinions and concerns. The following

documents provide a sampling of the issues raised in this critical debate. Religious as well as secular arguments are provided, as are the
guidelines issued by the National Bioethics Advisory Commission in
the United States.

DOCUMENT 125: Jon Cohen, "Can Cloning Help Save Beleaguered Species?" (1997)

One argument is that these new technologies can help preserve endangered species.

* * *

But Ryder [a geneticist from the San Diego Zoo's Center for Endangered Species] contends that the technology actually could be used to
increase the genetic diversity of a dwindling species. . . . Ryder also argues that cloning could be an especially useful tool for biologists trying
to save species that don't breed well in captivity, such as giant pandas.

Source: Jon Cohen, "Can Cloning Help Save Beleaguered Species?" *Science* 276
(May 30, 1997): 1329–1330.

DOCUMENT 126: Kirkpatrick Sales, "Ban Cloning? Not a Chance" (1997)

There are those who don't believe that cloning will or can be
banned. Sales, author of *Rebels against the Future: The Luddites and
Their War on the Industrial Revolution,* thinks that implementing technologies is almost inevitable.

* * *

But neither he [the President] nor Congress will be able to ban the
technological imperative that is inevitable in a culture built on the myth
of human power and the cult of progress. The essence of this imperative
was perhaps best defined by two men who crafted its apotheosis, the
atomic bomb.

"When you see something that is technically sweet you go ahead and
do it," said Robert Oppenheimer.

"Technological possibilities are irresistible to man," said John von
Neumann.

If the cloning of human embryos is possible—and no one really doubts anymore that it is—it will happen. In a world that not only permits but also commodifies gene-splicing, amniocentesis and in vitro fertilization, there cannot be any lasting legal restraints on any breakthrough in reproductive technology.

Source: Kirkpatrick Sales, "Ban Cloning? Not a Chance," *New York Times*, March 7, 1997, A31.

DOCUMENT 127: Council for Responsible Genetics, "Position Statement on Cloning" (1997)

The Council for Responsible Genetics, a private research group in Cambridge, Massachusetts, advised against human cloning.

* * *

I. We call upon the nations of the world to prohibit the cloning of human beings, by incorporating such prohibitions into their national laws and statutes.

II. We call upon the United Nations to take the initial steps by constituting an International Tribunal to articulate the concerns arising in different nations, cultures, religions and belief systems, with respect to the potential cloning of humans.

III. Within the United States, we call upon Congress to pass legislation to:

1. Prohibit the reproduction of human beings through processes other than development from the fertilized egg, and

2. to exclude animals and plants, their organs, tissues, cells or molecules from patenting, whether naturally occurring or cloned.

Source: Council for Responsible Genetics, "Position Statement on Cloning" (Cambridge, Mass.: Council for Responsible Genetics, 1997).

DOCUMENT 128: Pontifica Academia Pro Vita, "Reflections on Cloning" (1997)

The Vatican, through one of its research centers, presented its objections to cloning.

* * *

Human cloning belongs to the eugenics project and is thus subject to all the ethical and juridical observations that have amply condemned it.

It [cloning] represents a radical manipulation of the constitutive relationality and complementarity which is at the origin of human procreation in both its biological and strictly personal aspects. It tends to make bisexuality a purely functional left-over, given that an ovum must be used without its *nucleus in order to make room* for the clone-embryo and requires, for now, a female womb so that its development may be brought to term. This is how all the experimental procedures in zootechny are being conducted, thus changing the specific meaning of human reproduction.

In this vision we find the logic of industrial production: market research must be explored and promoted, experimentation refined, ever newer models produced.

Women are radically exploited and reduced to a few of their purely biological functions (providing ova and womb) and research looks to the possibility of constructing artificial wombs, the last step to fabricating human beings in the laboratory.

In the cloning process the basic relationships of the human person are perverted: filiation, consanguinity, kinship, parenthood. A woman can be the twin sister of her mother, lack a biological father and be the daughter of her grandfather. In vitro fertilization has already led to the confusion of parentage, but cloning will mean the radical rupture of these bonds.

Source: Pontifica Acaemica Pro Vita, "Reflections on Cloning" (Citta del Vaticano, Italy: Liberia Editrice Vaticana, 1997).

DOCUMENT 129: Constance Holden, "UN Weighs in on Cloning" (1997)

The United Nations created the Declaration on the Human Genome. The declaration consists of ethical guidelines for countries to follow.

* * *

In a flurry of moves in Europe addressing the potential worries surrounding human genome research, the Paris-based United Nations Educational, Scientific, and Cultural Organization (UNESCO) last week announced ethical guidelines to help members develop laws to address biotechnology's dizzying advances.

The "Declaration on the Human Genome" adopted unanimously by UNESCO's 186 member states outlines ethical research practices and calls for bans on any practice "contrary to human dignity," including human cloning.

Source: Constance Holden, "UN Weighs in on Cloning," *Science* 278 (November 21, 1997): 1407.

DOCUMENT 130: Katharine Q. Seeyle, "Clinton Bans Federal Money for Efforts to Clone Humans" (1997)

President Bill Clinton spoke out against cloning.

* * *

"Each human life is unique," Mr. Clinton said today, "born of a miracle that reaches beyond laboratory science. I believe we must respect this profound gift and resist the temptation to replicate ourselves."

Source: Katharine Q. Seeyle, "Clinton Bans Federal Money for Efforts to Clone Humans," *New York Times*, March 5, 1997, A13.

DOCUMENT 131: Munawar Ahmad Anees, "Human Cloning: An Atlantean Odyssey?" (1995)

Some members of the Islamic faith also questioned the morality of cloning, as represented in the following document.

* * *

In the Muslim consciousness, free from the inherent guilt, the body is, in a sense, an axis mundi. It is the point where the worlds, corporeal and spiritual, meet. . . . [T]he body is on trust from God. It is neither a solely owned nor a disposable commodity.

The Quranic paradigm of human creation, it would appear, preempts any move towards cloning. For the moment of birth to the point of death, the entire cycle is a Divine act. The humankind is simply an agent, a trustee of God, and the body a trust from God. As such, any replication is simply a redundant act. In the absence of a Quranic axiom on body as property, genetic interference in the germ-line would appear to be quite unethical.

On the utilitarian side of the corporeal possession, Muslims are exhorted to keep this gift given on trust in good shape. Given the case where cloning is an asexual experience (in the sense that it is performed within the legal marital bonds, no extramarital genetic boundaries are crossed, and the genetic endowment is only from the spouses), its prohibition must be judged against Islamic ethical norms. For instance, unlike Catholic doctrine, Islam sanctions therapeutic abortion in case of genuine clinical condition, e.g., impending danger to mother's life. Would cloning offer an analogous condition? We can think of only one possible scenario: prenatal corrective genetic intervention, provided there exists a clinical justification. Our reasoning for this assertion takes root in the body-as-a-trust paradigm and the ensuring responsibility for its care as the duty of every woman and man.

Source: Munawar Ahmad Anees, Editor-in-chief of *Periodica Islamica*, "Human Cloning: An Atlantean Odyssey?" *Eubios Journal of Asian and International Bioethics* 5 (1995): 36–37.

DOCUMENT 132: Rabbi Elliot Dorff, "Statement on Cloning" (1997)

Rabbi Elliot Dorff, a leading Jewish educator in Los Angeles, provided a Jewish perspective for the National Bioethics Advisory Committee.

* * *

Human cloning should be regulated, not banned. The Jewish demand that we do our best to provide healing makes it important that we take advantage of the promise of cloning to aid us in finding cures for a variety of diseases and in overcoming infertility. The dangers of cloning require that it be supervised and restricted. Cloning should be allowed only for medical research or therapy: the full and equal status of clones with other fetuses or human beings must be recognized with the equivalent protections guarded; and careful policies must be devised to determine how cloning mistakes will be identified and handled.

Source: Rabbi Elliot Dorff, Testimony to the National Bioethics Advisory Committee, March 14, 1997.

DOCUMENT 133: Dorothy C. Wertz, "Cloning Humans: Is It Ethical?" (1997)

At this point we can only speculate on the future effects of human cloning, and some people are not so sure cloning would be all bad.

* * *

The most likely use of cloning would be by people who cannot have children in the usual way, and who want to have a child who is like themselves, rather than inviting an unknown stranger's genes into the family. What is the harm in this?

Perhaps the strongest ethical argument against cloning is that it could lead to a new, untried type of family relationship. We have no idea what it would be like to grow up as the child of a parent who seems to know you "from inside," having gone through many of your own emotional crises in the same way.... On the other hand, most children want to have their own "secret places" where the parent cannot fathom what they think. The parent may impose expectations on the child, failing to realize that things may turn out very differently in a new historical period. Just because a family relationship is new and untried, is not a reason to condemn it automatically.... Parents and children may adjust to cloning far more easily than we think, just as has happened with in-vitro fertilization.

Source: Dorothy C. Wertz, "Cloning Humans: Is It Ethical?" *The Gene Letter* 1 (March 1997); found at http://www.geneletter.org.

DOCUMENT 134: Jerome P. Kassirer and Nadia A. Rosenthal, "Should Human Cloning Research Be Off Limits?" (1998)

Others can find important benefits from cloning.

* * *

Research on somatic-cell nuclear transfer might yield numerous benefits. Studies of stem-cell differentiation could provide valuable information about the mechanism of aging or the causes of cancer. Stem cells derived from this technology might also be a rich source of material for transplantation if specific genes or sets of genes in these pluripotent stem cells could be activated and if, as has been described before, the cells could then be coaxed to differentiate. Such a possibility is not strictly theoretical, because differentiated cell types . . . have been obtained by culturing embryonic stem cells from mice. If this technology could be applied to human stem cells, the resulting products might revolutionize medical therapeutics. The treatment of such diseases as diabetes mellitus, leukemia, and genetic disorders might change dramatically with the availability of genetically altered cell lines that would be immunologically compatible with a given patient and therefore not seen by the immune system as foreign.

The difficult ethical judgments about how to apply this new technology can be made only with full knowledge of the scientific facts. The burden of educating the public about these facts falls squarely on the shoulders of the scientists themselves, whose commitment to full disclosure may never be more stringently tested.

Source: Jerome P. Kassirer and Nadia A. Rosenthal, "Should Human Cloning Research Be Off Limits?" *New England Journal of Medicine* 338 (March 26, 1998): 905–906.

DOCUMENT 135: Lori B. Andrews, "Human Cloning: Assessing the Ethical and Legal Quandaries" (1998)

If we totally ban any and all cloning technologies, we could lose gains that can come from the careful application of this technology in specific areas.

* * *

Other proposals are too specific [for proposed legislation]. Some bills would prohibit the transfer of nucleic material from a somatic cell into a human egg. Not only could such provisions be evaded by using cow eggs, but they would also not prevent cloning via embryo splitting. In vitro fertilization clinics eventually want to be able to use this technology to enhance a woman's chances of becoming pregnant.

It is clear to me that neither the regulatory approach of the F.D.A. nor the narrow bans proposed by Congress and state legislatures are sufficient. I believe that, in line with the British model, we need to create a government oversight body with the authority to license fertility clinics, assess what reproductive technologies may be safely offered and by whom, and require the collection of follow-up data on the children created by these technologies.

Source: Lori B. Andrews, "Human Cloning: Assessing the Ethical and Legal Quandaries," *Chronicle of Higher Education*, February 13, 1998, B4–B5.

DOCUMENT 136: National Bioethics Advisory Commission, *Cloning Human Beings* (1997)

The National Bioethics Advisory Commission was commissioned to develop a set of recommendations for cloning.

* * *

I. The Commission concludes that at this time it is morally unacceptable for anyone in the public or private sector, whether in a research or clinical setting, to attempt to create a child using somatic cell nuclear transfer cloning. We have reached a consensus on this point because current scientific information indicates that this technique is not safe to use in humans at this point. Indeed, we believe it would violate important ethical obligations were clinicians or researchers to attempt to create a child using these particular technologies, which are likely to involve unacceptable risks to the fetus and/or potential child. Moreover, in addition to safety concerns, many other serious ethical concerns have been identified, which require much more widespread and careful public deliberation before this technology may be used.

The Commission, therefore, recommends the following for immediate action:

—A continuation of the current moratorium on the use of federal funding in support of any attempt to create a child by somatic cell nuclear transfer.

—An immediate request to all firms, clinicians, investigators, and professional societies in the private and non-federally funded sectors to comply voluntarily with the intent of the federal moratorium. Professionals and scientific societies should make clear that any attempt to create a child by somatic cell nuclear transfer and implantation into a woman's body would at this time be an irresponsible, unethical, and unprofessional act.

II. The Commission further recommends that:

—Federal legislation should be enacted to prohibit anyone from attempting, whether in a research or clinical setting, to create a child through somatic cell nuclear transfer cloning. It is critical, however, that such legislation include a sunset clause to ensure that Congress will review the issue after a specified time period (three to five years) in order to decide whether the prohibition continues to be needed. If state legislation is enacted, it should also contain such a sunset provision. Any such legislation or associated regulation also ought to require that at some point prior to the expiration of the sunset period, an appropriate oversight body will evaluate and report on the current status of somatic cell nuclear transfer technology and on the ethical and social issues that its potential use to create human beings would raise in light of public understandings at that time.

III. The Commission also concludes that:
—Any regulatory or legislative actions undertaken to effect the forgoing prohibition on creating a child by somatic cell nuclear transfer should be carefully written so as not to interfere with other important areas of scientific research. In particular, no new regulations are required regarding the cloning of human DNA sequences and cell lines, since neither activity raises the scientific and ethical issues that arise from the attempt to create children through somatic cell nuclear transfer, and these fields of research have already provided important scientific and biomedical advances. Likewise, research on cloning animals by somatic cell nuclear transfer does not raise the issues implicated in attempting to use this technique for human cloning, and its continuation would only be subject to existing regulations regarding the humane use of animals and review by institution-based animal protection committees.
—If a legislative ban is not enacted or if a legislative ban is ever lifted, clinical use of somatic cell nuclear transfer techniques to create a child should be preceded by research trials that are governed by the twin protections of independent review and informed consent, consistent with existing norms of human subjects protection.
—The United States Government should cooperate with other nations and international organizations to enforce any common aspects of their respective policies on the cloning of human beings.

IV. The Commission also concludes that different ethical and religious perspectives and traditions are divided on many of the important moral issues that surround any attempt to create a child using somatic cell nuclear transfer techniques. Therefore, we recommend that:
—The federal government, and all interested and concerned parties, encourage widespread and continuing deliberation on these issues in order to further our understanding of the ethical and social implications of this technology and to enable society to produce appropriate long-

term policies regarding this technology should the time come when present concerns about safety have been addressed.

V. Finally, because scientific knowledge is essential for all citizens to participate in a full and informed fashion in the governance of our complex society, the Commission recommends that:
    —Federal departments and agencies concerned with science would cooperate in seeking out and supporting opportunities to provide information and education to the public in the area of genetics, and on other developments in the biomedical sciences, especially where these affect important cultural practices, values, and beliefs.

Source: National Bioethics Advisory Commission, *Cloning Human Beings* (Washington, D.C.: U.S. Government Printing Office, 1997).

Selected Bibliography

GENERAL

For a convenient overview of issues related to genetic engineering, confer the various articles in the 107-page entry on genetics in *The Encyclopedia of Bioethics*, revised edition, ed. Waren Reich (New York: Macmillan Library Reference. 1995), vol. 2, pp. 907–1020. Each individual article has cross-listings to related articles and bibliography with it.

Special Supplement "Genetic Grammar: 'Health,' 'Illness,' and the Human Genome Project" in *The Hastings Center Report* 22 (July–August 1992): S11–S20.

JOURNALS

Alper, J. "Genetic Testing and Insurance." *British Medical Journal* 307 (December 1993): 1506–1507.

Anderson, W. French. "Genetic Engineering and Our Humanness." *Human Gene Therapy* 5 (1994): 755–759.

———, "Genetics and Human Malleability." *Hastings Center Report* 20 (1990): 21–24.

———. "Human Gene Therapy: Why Draw a Line?" *Journal of Medicine and Philosophy* 14 (December 1989): 681–693.

Anderson, W. French, and T. Friedmann. "Strategies for Gene Therapy." In *The Encyclopedia of Bioethics*, revised edition, ed. Waren Reich. New York: Macmillian Library Reference, 1995. Vol. 2, p. 908.

Andrews, Lori B. "Human Cloning: Assessing the Ethical and Legal Quandaries." *Chronicle of Higher Education*, 13 February 1998, B4–B5.

Angier, Natalie. "Ultrasound and Fury: One Mother's Ordeal." *New York Times*, 26 November 1996, C1 and C8.

Berg, Paul et al. "Asilomar Conference on Recombinant DNA Molecules." *Science* 188 (June 1975): 991–994.

Berger, Edward M. "Morally Relevant Features of Genetic Maladies and Genetic Testing." In Bernard Gert et al., *Morality and the New Genetics*. Sudbury, Mass.: Jones and Bartlett, 1996.

Blatt, Robin J. R. "An Overview of Genetic Screening and Diagnostic Tests in Health Care." *Gene Letter* 1 (September 1996), an electronic journal found at http://www.geneletter.org.

Bonnicksen, Andrea L. "Ethical and Policy Issues in Human Embryo Twinning." *Cambridge Quarterly of Healthcare Ethics* 4 (1995): 268–284.

Brumby, Margaret, and Pascal Kasimba. "When Is Cloning Lawful?" *Journal of In Vitro Fertilization and Embryo Transfer* 4 (August 1987): 198–204.

Cohen, Stanley. "The Manipulation of Genes." *Scientific American* 233 (July 1975): 25–33.

Collins, Francis S. "BRCA1—Lots of Mutations, Lots of Dilemmas." *New England Journal of Medicine* 334 (January 1996): 186–88.

Cordi, A. M., and J. Brandt. "Psychological Cost and Benefits of Predictive Testing for Huntington's Disease." *American Journal of Medical Genetics* 55 (1995): 618–625.

deWachter, Maurice A. M. "Ethical Aspects of Human Germ-Line Therapy." *Bioethics* 7 (1993): 166–177.

Dietrich, Donald J. "Catholic Eugenics in Germany, 1920–1945: Hermann Muckermann, S. J. and Joseph Mayer." *Journal of Church and State* 34 (summer 1992): 575–600.

Friedman, J. M. "Eugenics and the 'New Genetics.'" *Perspectives in Biology and Medicine* 35 (autumn 1991): 145–154.

Gaylin, Willard. "The Frankenstein Myth Becomes a Reality: We Have the Awful Knowledge to Make Exact Copies of Human Beings." *New York Times Magazine*, March 5, 1972.

Grobstein, Clifford, and Michael Flower. "Gene Therapy: Proceed with Caution." *Hastings Center Report* 14 (April 1984): 13–17.

Groopman, Jerome, M.D. "Decoding Destiny." *New Yorker* 76 (February 1998): 42–48.

Handyside, Allan H., et al. "Birth of a Normal Girl after In Vitro Fertilization and Preimplantation Diagnostic Testing for Cystic Fibrosis." *New England Journal of Medicine* 327 (September 1992): 905–909.

The Hastings Center. "Cloning Human Beings: Responding to the National Bioethics Advisory Commission's Report." In *Hastings Center Report* 27 (September–October 1997): 6–22.

Healy, Bernadine. "BRCA Genes—Bookmaking, Fortunetelling, and Medical Care." *New England Journal of Medicine* 336 (May 1997): 1448–1449.

Hubbard, Ruth, and R. C. Lewontin. "Pitfalls of Genetic Testing." *New England Journal of Medicine* 334 (May 1996): 1192–1193.

Juengst, Eric. "Germ-Line Therapy: Back to the Basics." *Journal of Medicine and Philosophy* 16 (1991): 587–592.

Kass, Nancy. "Insurance for the Insurers: The Use of Genetic Tests." *Hastings Center Report* 22 (November–December 1992): 6–11.

Keenan, James, S. J. "Genetic Research and the Elusive Body." In L. S. Cahill and M. A. Farley, eds., *Embodiment, Morality, and Medicine*. The Netherlands: Kluwer Academic Publishers, 1995, pp. 59–73.

Kessler, David A., et al. "Regulation of Somatic Cell Therapy and Gene Therapy by the Food and Drug Administration." *New England Journal of Medicine* 329 (October 1993): 1169–1173.

Kolata, Gina. "Clinics Selling Embryos Made for 'Adoption.' " *New York Times*, November 23, 1997, A1 and A18.

———. "Scientists Brace for Changes in Path of Human Evolution." *New York Times*, March 21, 1998, A1 and A7.

Leiden, Jeffrey M. "Gene Therapy—Promise, Pitfalls, and Prognosis." *New England Journal of Medicine* 333 (September 1995): 871–72.

Lewontin, R. C. "The Dream of the Human Genome." *New York Review of Books*, May 28, 1992, 31–40.

Lui, J., et al. "Birth after preimplantation diagnosis of the cystic fibrosis delta F508 mutation by polymerase chain reaction in human embryos resulting from intracytoplasmic sperm injection with epididymal sperm." *Journal of the American Medical Association* 272 (December 1994): 1858–1860.

Moreno, Jonathan D. "Private Genes and Public Ethics." *Hastings Center Report* 13 (October 1983): 5–6.

Muller, Hermann J. "Human Evolution by Voluntary Choice of Germ Plasm." *Science* 134 (September 1961): 643–649.

Murray, Mary. "Nancy Wexler." *New York Times Magazine*, February 13, 1994, 31.

Murray, Thomas H. "Genetics and the Moral Mission of Health Insurance." *Hastings Center Report* 22 (November–December 1992): 12–17.

———. "Warning: Screening Workers for Genetic Risk." *Hastings Center Report* 13 (February 1983): 5–8.

Post, S. G. "Genetics, Ethics and Alzheimer's Disease." *Journal of the American Geriatric Society* 42 (July 1994): 782–786.

Reilly, Philip R. "Genetic Privacy Bills Proliferate." *Gene Letter* 1 (May 1997) http://www.geneletter.org.

Rennie, J. "Grading the Gene Tests." *Scientific American* (June 1995): 88–97.

Rosenfeld, Albert. "At Risk for Huntington's Disease: Who Should Know What and When?" *Hastings Center Report* 14 (June 1984): 5–8.

Schrag, Deborah, et al. "Decision Analysis—Effects of Prophylactic Mastectomy and Oophorectomy on Life Expectancy among Women with BRCA1 and BRCA2 Mutations." *New England Journal of Medicine* 336 (May 1997): 1465–1471.

Sendelowski, M. "Separate, but Less Equal: Fetal Ultrasonography and the Transformation of Expectant Mother/Fatherhood." *Gender and Society* (1994) 8: 230–43.

Shickle, D., and R. Chadwick. "The Ethics of Screening: Is 'Screeningitis' an Incurable Disease?" *Journal of Medical Ethics* 20 (1994): 12–18.

Simpson, Joe, and Sandra Carson. "Preimplantation Genetic Diagnosis." *New England Journal of Medicine* 327 (September 1992): 951–953.

Vacek, Edward, S. J. "Vatican Instruction on Reproductive Technology." *Theological Studies* 49 (March 1988): 110–131.

Wade, Nicholas. "Recombinant DNA: NIH Sets Strict Rules to Launch New Technology." *Science* 190 (December 1975): 1175–1179.

Walters, LeRoy. "Human Gene Therapy: Ethics and Public Policy." *Human Gene Therapy* 2 (summer 1991): 116–120.

Watson, James D. "The Human Genome Project: Past, Present, and Future." *Science* 248 (April 1990): 44–48.

———. "Moving toward Clonal Man: Is This What We Want?" *The Atlantic*, May 1971, 50–53.

Wertz, Dorothy C., et al. "Genetic Testing for Children and Adolescents: Who Decides?" *Journal of the American Medical Association* 272 (September 1994): 875–881.

Wiggins, Sandi, et al. "The Psychological Consequences of Predictive Testing for Huntington's Disease." *New England Journal of Medicine* 327 (November 1992): 1401–1405.

Wivel, Nelson A., and LeRoy Walters. "Germ-Line Gene Modification and Disease Prevention: Some Medical and Ethical Perspectives." *Science* 262 (October 1993): 533–538.

Zimmerman, Burke K. "Human Germ-Line Therapy: The Case for Its Development and Use." *Journal of Medicine and Philosophy* 16 (1991): 593–612.

BOOKS

Basen, Gwynne, Margrit Eicher, and Abby Lippman, eds. *Misconceptions: The Social Construction of Choice and the New Reproductive and Genetic Technologies*. Quebec: Voyageur Publishing, 1996.

Carmen, Ira H. *Cloning and the Constitution: An Inquiry into Governmental Policymaking and Genetic Experimentation*. Madison: University of Wisconsin Press, 1985.

Catholic Health Association of the United States. *Human Genetics: Ethical Issues in Genetic Testing, Counseling, and Therapy*. St. Louis, Mo.: Catholic Health Association of the United States, 1990.

Congregation for the Doctrine of Faith. *Donum Vitae*. Instruction on Respect for Human Life in Its Origin and on the Dignity of Procreation. February 22, 1987. In Thomas A. Shannon and Lisa S. Cahill, *Religion and Artificial Reproduction*. New York: Crossroad, 1988.

Ebon, Martin, ed. *The Cloning of Man: A Brave New Hope—or Horror?* New York: New American Library, 1978.

Gregg, R. *Pregnancy in a High-Tech Age: Paradoxes of Choice*. New York: Paragon House, 1993.

Kevles, Daniel J. *In the Name of Eugenics: Genetics and the Uses of Human Heredity*. New York: Knopf, 1985.

Kevles, Daniel J., and LeRoy Hood. *The Code of Codes: Scientific and Social Issues in the Human Genome Project*. Cambridge: Harvard University Press, 1992.

Kolata, Gina. *Clone: The Road to Dolly and the Path Ahead*. New York: William Morrow, 1998.

Kuehl, Stefan. *The Nazi Connection: Eugenics, American Racism and German National Socialism*. New York: Oxford University Press, 1994.

Lewontin, R. C. *Biology as Ideology: The Doctrine of DNA*. New York: Harper Perennial and Harper Torchbooks, 1991.

Lyon, Jeff, and Peter Gorner. *Altered Fates: Gene Therapy and the Retooling of Human Life*. New York: W. W. Norton, 1995.

Marteau, Theresa, and Martin Richards, eds. *The Troubled Helix: Social and Psychological Implications of the New Human Genetics*. Cambridge: Cambridge University Press, 1996.

McGee, Glenn. *The Perfect Baby: A Pragmatic Approach to Genetics* Lanham, Md.: Rowman and Littlefield, 1997.

Muller, Hermann J. *Out of the Night: A Biologist's View of the Future*. New York: Garland, 1984 (c. 1935).

National Bioethics Advisory Commission (NBAC). *Cloning Human Beings: Report and Recommendation of the National Bioethics Advisory Commission*. Rockville, Md.: NBAC, 1997.

Nelkin, Dorothy, and M. Susan Lindee. *The DNA Mystique: The Gene as Cultural Icon*. New York: W. H. Freeman, 1995.

Pence, Gregory E. *Who's Afraid of Human Cloning?* Lanham, Md.: Rowman and Littlefield, 1998.

President's Commission for the Study of Ethical Problems in Medicine and Biomedical and Behavioral Research. *Screening and Counseling for Genetic Conditions*. Washington, D.C.: U.S. Government Printing Office, 1983.

———. *Splicing Life: The Social and Ethical Issues of Genetic Engineering with Human Beings*. Washington, D.C.: U.S. Government Printing Office, 1982.

Ramsey, Paul. *Fabricated Man: The Ethics of Genetic Control*. New Haven, Conn.: Yale University Press, 1970.

Rothman, Barbara Katz. *The Tentative Pregnancy*. New York: Viking, 1986.

Shannon, Thomas A., and Lisa S. Cahill. *Religion and Reproduction*. New York: Crossroad, 1988.

Thompson, Larry. *Correcting the Code: Inventing the Genetic Cure for the Human Body*. New York: Simon and Schuster, 1994.

Verlinshy, Y., and A. M. Huliev, eds. *Preimplantation Diagnosis of Genetic Diseases: A New Technique in Assisted Reproduction*. New York: Wiley-Liss, 1995.

Walters, LeRoy, and Julie Gage Palmer. *The Ethics of Human Gene Therapy*. New York: Oxford University Press, 1997.

Watson, James D., and John Tooze. *The DNA Story: A Documentary History of Gene Cloning*. San Francisco: W. H. Freeman, 1981.

Index

Abnormality, 108–109
AIDS, 32, 170, 177, 242
American Society of Human Genetics,
    150, 151, 166
Anderson, W. French, 136, 197
Andrews, Lori, 269
Anees, Munwar, 266
Angier, Natalie, 197
Annas, George, 191
Animal Welfare Act, 45
Animals, transgenic, 39–40; use in re-
    search, 40, 45, 53–56, 235
Asgrow Seed Co. v. Winterboes, 91
Asilomar Conference, xxvi, 4, 6, 8–10,
    13, 18, 230, 240

Baltimore, David, 8
Barinaga, Marcia, 195
Beckwith, Jonathan, 33
Berg, Paul, 5–6, 8, 11
Betsch, David, 183
Biodiversity, 44
Biodiversity: convention on, 83, 179
Bioethics, definition of, 149
Biological piracy, 250
Boyd, Robert, 186
BRCA1, 151, 216
Breast cancer: insurance, 151, 210;
    testing, 151, 210
Brenner, S., 8

Brody, Baruch, 56
Bruce, Donald, 170

Callahan, Daniel, 10
Catholic Health Association, 137
Cloning: animals, xxix, xxx, 35, 235,
    266; benefits, 268–269; DNA, xxv,
    xvi–xxx, 7, 35, 100; ethics, 261, 256–
    270; legislation, 259, 263–264, 266,
    269–68, 270–272; prohibition, 264,
    265, 266, religion, 264–267; reproduc-
    tion, 262, 268; technology, 259–260,
    262; verification, 262
Cohen, John, 263
Cohen, Stanley, xxvii, 20
Cole-Turner, Ron, 123
Confidentiality, 111, 115, 168, 191
Cook-Deegan, Robert, 104, 138, 140
Council for Responsible Genetics, 264
Created co-creator, 121–122
Crick, Francis, xxv, 3, 30, 130
Crop: diversity, 8; genetic engineering;
    80
Cystic Fibrosis, 150, 206

Danish Ethics Council on Animals, 43
Department of Energy, 102, 113, 138,
    243
Determinism, biological, 172
Diamond v. Chakrabarty, 57, 245–249

Disclosure, 114–115, 168
Discrimination, 121, 131, 160, 166, 214, 219
DNA: data banks, 34, 116–117; human rights, 187; identification, 183, 185–186, 189; quality control, 184–185; sanctity of, 123; structure, xxv, 3, 30, 44
Dolly, 259, 260, 261
Donis-Keller, Helen, 99
Donum Vitae, 127
Dor Yeshorim, 205
Dorff, Elliot, 267
The Double Helix, xxv, 3, 27, 29
Dulbecco, Renato, 100

Eadie, Michelle, 183
Enhancement, 164, 168, 170, 199, 233, 236, 253
EPA guidelines, 73–75; diversity, 88; environmental risks, 87–90; Europe, 85–87; 72–73; genetic engineering, 64–65; regulations, 66, 72–73, 75, 86, 90; safety of, 65–72
Episcopal Church, 125
Eugenics, xxiv, 233, 242

Feder, Barnaby, 82
Food and Drug Administration (FDA), 66, 71, 72
Food, genetically engineered, xxvi–xviii; against nature, 64; allergenic reaction, 69, 85; benefits, 66, 79–82; commercialization, 88; costs, 83; critique of EPA and transgenic food, 73–76; development of, 79–81; diversity, 89; drugs, to produce, 78; ecological risks, 87–90; European reaction, 85–87; public perception, 65; regulation, 66–67, 71–72, 73; rice, 78–82
Francione, Gary, 45
Frankenfood, 64
Frankenstein Factor, 227–232
Franklin, Rosalin, 30

Garrod, Archibald, xxiv
Gaylin, Willard, 227

Gene therapy, 34, 39, 124, 126, 133, 136, 137, 169, 195, 198, 200, 222; ethics, 162, 163, 171, 198, 202; germ line, 34, 39, 161–162, 171, 232; rDNA Advisory Board Guidelines, 169; research, 222–223; safety, 65, 71, 89, 198, 254–255; somatic-cell, 169
Genetic commodification, 182
Genetic Confidentiality and Nondiscrimination Act, 115
Genetic engineering: control of, 7; economic issues, 85, 88, 91–96, 108; employment, 115–116, 152, 157, 158–159, 159–161; environment, 152; ethics, 167, 230–232, 246–243, 253–255; social implications, 7
Genetic information: family, 187; NIH-DOE policy, 245; protection for, 244
Genetic Information Nondiscrimination in Health Insurance Act of 1997, 114
Genetic Justice Act, 115
Genetic monitoring in the workplace, 155
Genetic research: for enhancement, 168; on human embryo, 126; on genome, 149
Genetic screening, 108–109, 112, 113, 168, 206, 237; difficulties with, 242; disease, 176–177; ethics, 253; insurance, 244–245; Office of Technological Assessment, 151–160; politics, 150–151; presymptomatic diagnosis, 244; prohibition, 156; workplace, 154
Genetic testing, 211–213; applications, 201, 205–209, 241; criteria for, 113; discrimination, 210–214; ethics, 198, 212; insurance, 114, 166, 210, 212–213, 214–220, 242, 245; marriage, 204–209; psychological, 202–203; reproductive, 112, 205–209; workplace, 153, 154–160
Greely, Henry, 174
Green, Howard, xxvii

Haldane, J.B.S., 151
Harvey, William, xxiii
Hayes, Catherine, 200

Hefner, Philip, 119, 121
Holden, Constance, 265
Hudson, Kathy, 214
Human embryo research, 126–127
Human Genome Organization
    (HUGO), 139, 143, 187
Human Genome Project, xxix, 28, 31,
    102, 104–105, 110–111, 128, 139, 191,
    238; abortion, 132; benefits, 104–105,
    110–111, 124, 140–142; commerciali-
    zaton, 175, 177, 251; control of gene-
    tic information, 182; discrimination,
    114–115, 116, 117, 121, 131–132, 160,
    166; ethical issues, 120–121, 134,
    144, 149, 238; human nature, 107;
    image of self, 107; normal, 108; pat-
    enting policy, 128–130; religion, 105–
    107, 118–120
Human growth factor, 199, 236
Human being, 58, 107, 118, 127, 134,
    232
Huntington's Disease, xxix, 109, 200–
    205, 210, 215

Ibrahim, Youssef, 84
Indigenous people, 175–177, 179, 180,
    182,
Insurance, 114, 114–115, 161, 163, 214,
    216–217, 242
Isner, Jeffrey, 200

Jackson, Joe, 186
Jaenish, Rudolf, 39

Kass, Leon, 57
Kassirer, Jerome, 268
Keenan, James, 232
Kegan, Jerome, 171
Kessler, David, 65
King, Mary-Claire, 187
King, Patricia, 33
Knock-out genetic techniques, 41
Kohr, Martin, 250
Kolata, Gina, 185, 202, 205, 261, 262
Koshland, Daniel, 34, 110

Langan, Michael, 169
Leder, Philip, 46

Leiden, Jeffrey, 221
Liloqula, Ruth, 182

Maddox, John, 240
Magel, Charles, 251
Map, genetic, 94, 99, 100–101, 103,
    141, 144; benefits, 104, 141, 142
McCouch, Susan, 78–82
Mendel, Gregor, xxiv
Mice, knock-out, 41–43
Milk, enhanced production, 63
Miller, Henry, 73
Moffat, Anne Simon, 77
Morgan, Thomas, xxv
Mudd, James, 184
Mullens, Kerry, xxix
Mulligan, Richard, 195
Murray, Robert, 33
Murray, Thomas, 149, 212

National Bioethics Advisory Commis-
    sion, 270
National Council of Churches, 59, 123–
    124
National Institutes of Health, 1, 9, 10,
    11, 15, 16, 27, 112, 113, 130, 168, 243;
    DNA guidelines, xxvi–xxvii
Normal, idea of, 108, 168, 199

Office of Technology Assessment, 152,
    154
O'Neill, Molly, 64

Painter, Thomas, xxv
Patenting: animals, 46, 47–51, 52–53, 57–
    60, 251; crops, 91–96; economic im-
    plications; 52; engineered bacteria,
    246–249; human applications, 49, 178;
    Human Genomic data, 128, 174–181;
    religion, 52, 57–60
Pauling, Linus, 30
Peters, Ted, 117, 130
Plant engineering, environmental im-
    pact, 83–84
Plato, xxii, xxiv
Playing God, 105, 125, 199, 254
Polymerease Chain Reaction (PCR),
    xxix

Pontifica Academia Pro Vita, 264
Prenatal diagnosis, 127, 236–238
President's Commission for the Study
  of Ethical Problems in Medicine
  and Biomedical and Behavioral Research, 167, 168, 252
Presymptomatic diagnosis, 151, 200,
  243

Recombinant DNA (rDNA); Asilomar
  guidelines, 9; benefits, 23–24; fears
  of, 228–229, 230; industry, 26–27;
  NIH guidelines, risks, 15, 19–20, 21–
  23, 23; research, xxvi, xxv, 1, 5, 9–
  10, 15, 21, 27, 66, 75, 165, 195, 227;
  research guidelines, 11, 16–18, 18–
  19, 20, 21–23; safety, xxv, 5–6, 8, 10,
  12, 14, 15, 230; technology of, 1, 5,
  21
Repository for Germinal Choice, 236
Restriction Fragment Length Polymor-
  phisms (RFLPS), xxvii
Rifken, Jeremy, 20, 26, 47, 52, 112,
  122
Rights, animal, 43, 45, 47–51, 251
Rights, human, 6, 178, 178–181, 187–
  190
Rissler, Jane, 87
Roblin, R., 8
Rosenberg, S., 196
Rosenthal, Nadia, 268
Roslin Institute, xxix
Rothstein, Mark, 112
Rural Advancement Foundation Inter-
  national (RAFI), 178, 182
Russell, Robert, 117

Sagoff, Mark, 51, 251
Sales, Kirkpatrick, 263
Sanctity of life, 58, 134
Seele, Katharine, 266
Sgaramella, Vittorio, 260
Shannon, Thomas, 234
Shuldiner, Alan, 53

Silverstone, Allen, 19
Simpson, O. J., 185, 186
Singer, Maxine, 8
Sinsheimer, Robert, 4, 138
Specter, Michael, 261
Stewardship, 59–60, 125
Stewart, Timothy, 46
Stocel, Abadio, 182

Tarasoff v. Regents of the University of
  California, 111
Tauer, Carol, 103
Traits: dominant and recessive, xxiv;
  sex-linked, xxv
Transgenic animals, 39, 43, 44, 46, 47,
  49, 53, 57–60, 235
Travis, John, 41

Union of Concerned Scientists, 90
United Church of Christ, 125
United Methodist Church, 124, 126

Wachbroit, Robert, 47
Wade, Nicholas, 12, 18
Walters, LeRoy, 161
Watson, James, xxv, 3, 18, 27–35, 103,
  104, 118, 130, 138, 140, 259
Weber, Barbara, 210
Weiss, Mary, xxviii
Weissman, August, xxiv
Wertz, Dorothy, 268
Wexler, Nancy, 33
Wheeler, David, 78
Wilkner, Nachama, 116
Wills, Peter, 44, 63
Wilmot, Ian, xxix, 259, 261
Wivel, Nelson, 161
Wolff, Caspar, xxiii
World Council of Churches (WCC),
  126, 137

Yanagimachi, Ryuzo, xxx

Zinder, Norton, 260

About the Editor

THOMAS A. SHANNON is Professor of Religion and Social Ethics in the Department of Humanities and Arts at Worcester Polytechnic Institute. He is the author and editor of many books and articles on bioethics and Catholic social teaching.